The Forest Rangers

The FOREST RANGERS

Montgomery M. Atwater

Philadelphia
MACRAE SMITH COMPANY

To my sister, Betty,
in love and admiration

The author and editors appreciate the help of the Forest Service in reviewing this book and providing many of the photographs.

SBN: 8255-1346-4
Library of Congress Catalog Card Number 74-87989
Manufactured in the United States of America
Book design by Adrianne Onderdonk Dudden

6910

Second Printing

Contents

Author's Foreword

The Forest Rangers is not a formal history of the United States Forest Service. Such a history would be a worthwhile project. But I believe that I have undertaken something just as important and possibly more interesting in seeking to capture for the reader the spirit and the character of this great organization through the deeds and personalities of the men who made it, and still are making it, what it is.

I was a forester myself for eighteen years and have lived and worked among foresters for at least another ten. Therefore no apology is offered for the fact that I am occasionally a part-time actor as well as narrator in this story. Where men I have known and adventures in which I have taken part personally seemed to tell the story best, I have used them. Firsthand experience is always the most vivid.

It is my hope that this book will inspire some of its readers to become foresters themselves—girls as well as boys. To both sexes the Forest Service offers lifetime careers in which no one is likely to get wealthy but no one will ever be bored. For the benefit of anyone who does have an ambition to become a forester, there is a chapter at the end of this book telling how to go about it.

M.M.A.

1

Ordeal by Fire

It was a hot summer. In the deep canyons, and over the forest-clad peaks of the Bitterroot Mountains, heat lay like a blanket, day after day. The wildflowers bloomed early and wilted. In the alpine meadows the lush grasses turned from green to brown. The smell of pitch was strong in the aisles of spruce and fir and pine; the color of their needles became a hard bluish-green; and the inches-deep carpet of needles, leaves and dead grass crackled underfoot. The oldest inhabitants could remember nothing like it. They were afraid, and they had good reason for their fear —forest fire.

Fire, controlled, is one of man's best friends. By means of fire he lifted himself above the level of the beasts. It cooks his food, warms his home, and makes the steam and electricity that power industry. Uncontrolled, fire is the most horribly destructive of the natural forces, for it feeds and grows stronger on what it destroys. In the Northwest, in that summer of 1910, a forest area two hundred miles long by seventy-five miles wide had become a heap of tinder waiting for the match.

In those days, much of this region of northeastern Washington, northern Idaho and western Montana was virtual frontier. Two railroads, one of them still under construction, traversed it from east to west. There were a few wagon roads and horse trails. Settlements were strung out along the railroads like a very few beads on a very long chain. The largest of them, Wallace, Idaho, in the heart of the Bitterroots, had only thirty-five hundred inhabitants. The others, with names soon to be printed in scare headlines all over the world, were mining towns, lumber mills and railroad construction camps.

The Bitterroot Mountains run from near the Canadian border south for two hundred miles almost midway between the Cascades of Washington and the Rockies of Montana. It is not a lofty range; the tallest peaks are in the seven-thousand-foot class. But it is very rugged. It was then and much of it still is a forest primeval. In 1910 most of it was not even explored. The Indians never penetrated beyond the fringes. They couldn't cope with the dense tangle of trees and brush, the wild rivers and snowbound passes. Palefaces hadn't done much better. Lewis and Clark fought their way through it. A few trappers, loggers, miners and homesteaders had inched their way through the canyons. There were no paved highways, no airports, no radio stations and very few telephone lines. By mid-August of 1910, hundreds of lightning-set fires were already burning in the Bitterroots. This, then, was the stage setting for what was to become known as the Great Idaho Fire.

At a settlement in Burke Canyon, a few miles from Wallace, Forest Ranger Edward C. Pulaski had recently taken over his duties in the Coeur d'Alene National Forest. There had been no ceremony. There was a letter from Forest Supervisor Weigle directing him to administer his district. The Forest Service would pay him sixty dollars per month. He would furnish his own horse.

In two sentences, Pulaski accepted sole responsibility for the welfare of a wilderness empire the size of a county —a big one. There were not very many people in it. He probably knew most of them by name. On the other hand, he had none of the trappings of a modern administrator—no office, no assistants, no fleet of cars and trucks or roads to drive them on, no communications network. He had himself and his horse. At that time and in that place he *was* the Forest Service, a federal bureau that had just been established by President Theodore Roosevelt and was just beginning to get organized.

It was a time when many new national forests were taking shape in the West by presidential order, twenty-two of them containing sixty-four thousand square miles in a single block, with the Bitterroots as its hub. Many Forest Rangers were assuming their great obligations as informally as Pulaski. There was one difference. Destiny had fixed its eye on Big Ed, as he was known throughout northern Idaho. He was about to be tested, and the Forest Service with him, in an ordeal by fire.

Big Ed Pulaski was forty years old when he became a Forest Ranger. At that age most men are settled in their jobs and are even thinking about taking life a little easier instead of embarking on a new and difficult career. He was not a college graduate and might have had trouble writing a schoolbook definition of the words "administer" and "conservation." On the other hand, he understood the practical meaning of those words very well. He was a graduate of the hard school of the frontier, with twenty years' experience as packer, prospector, logger and woodsman. He knew his way around the mountains as a city dweller knows the way around his own block.

He had helped build the few roads. He had followed the dim trails made by man until they petered out, and the dimmer trails made by deer and elk until they petered out too, and then he had found his own way. In nameless canyons he had camped by nameless streams. He had looked upon forests of noble white pine measured in square miles. Beneath the rainproof canopies of cedar and spruce he had taken shelter from the violent mountain storms. He knew by what secret passes the elk migrated from the high country to their winter feeding grounds, how the beaver built their dams, and where the fish went to spawn.

In a land where people were few he had already seen the waste that thoughtless men can create—whole forests deliberately burned just to uncover a vein of ore. He had rebelled against this waste. He was ready-made by training and desire for the Forest Service as the Forest Service was ready-made for him. But certainly in that ominous summer of 1910, Pulaski had no idea that he was about to become the first hero of the Forest Service.

In all the scattered settlements and camps of the Bitterroots, people were living with a constant reminder of peril —the acrid odor of burning and a pall of smoke so dense that it turned noonday into dusk. At night, that smoke pall glowed red from horizon to horizon. Along the Coeur d'Alene River, which with its tributaries was the center of Pulaski's district, normal business had ceased. Practically all able-bodied residents, and as many more as could be recruited from distant cities like Butte, Montana, and Spokane, Washington, were fighting fire.

Some families had already evacuated, for in these wilderness settlements the forest came right to the doorsteps of their cabins. By packhorse, wagon or railroad they took with them what possessions they could, having buried the rest. But many had stayed, unwilling to desert their men or surrender their homes without a battle. Pulaski's wife, Emma, was one who refused to go.

On August 19, Big Ed came in from the firelines for supplies and to give his wife some final instructions. As he knew better than anyone, the situation was getting worse by the hour. New fires were breaking out every day; old ones were getting bigger. If all these scattered blazes should meet, the result would be holocaust. If fire came down Burke Canyon, he told Emma Pulaski, she must not attempt to flee to Wallace. That town, he thought, would surely burn. Emma was to take refuge on a nearby tailings dump, a mass of waste from the mines which made a large gap in the forest.

When Emma begged him not to go back into the woods, the ranger shrugged his big shoulders. He had men to look after. "I can always take care of myself," he said, "unless the whole shebang blows up."

At that moment a message came from Supervisor Weigle that a new fire had sprung up on Placer Creek. A crew was on the way. Would Pulaski take charge of it and arrange for equipment and supplies? When he vanished into the smoke, Emma Pulaski wondered if she would ever see her husband again.

Big Ed was in overall charge of several crews, each fighting a separate fire. He deployed them around the perimeter. Doggedly they cut brush and trees and scraped away leaves and grass to make a trench across which ground fire would not creep. They were weary now, from days of this work in heat and smoke, and they were despondent from the seeming futility of it. They couldn't put all these fires out, or even slow them down much with their puny trenches. Trenches and firebreaks were effective at the right time and place, but not against a monster that could jump a river.

The uneven struggle was not futile to Pulaski. He understood that they could not put out the fires. But if they could keep the fires separated—prevent them from turning the area into one single torrent of flame—perhaps men could hold out until the fall rains. Perhaps they could save the towns, their families—and themselves.

In Missoula, Montana, two men were thinking along somewhat the same lines. Missoula was headquarters of Region One of the Forest Service, which included all the national forests between eastern Washington and western South Dakota. The men were Regional Forester William

B. Greeley and his assistant, Ferdinand A. Silcox. (In 1910, Greeley's correct title was District Supervisor of District One. I have used the modern terminology in order to avoid confusion). Each of these men was a future chief of the Forest Service. At the moment, their only concern was the desperate plight of the Coeur d'Alene National Forest, where the fires were on the edge of overwhelming the stubborn defenders. They understood the situation well, for they had just come in from the firelines themselves.

There was some slight reason for optimism. President Taft had just authorized the use of troops for fire fighting. This would mean manpower to relieve and reinforce the exhausted crews. While the weather held as hot and dry as ever, there was little wind. The fires were spreading slowly. If those rangers in Idaho could hang on a few more days, the weather *must* break. The fire-killing rains would come.

But the next day, August 20, at about four in the afternoon, the whole shebang did blow up.

Today, the mechanics of a fire storm are better understood. Any fire creates a draft. It heats the air and consumes part of it. As the hot gases rise, more air flows in from the sides to fill the partial vacuum. There is a self-sustaining chain reaction as long as there is fuel to feed the flames. If the fire gets hot enough and big enough, these air currents generated by the fire itself become violent. The bombing of the German city of Dresden in World War II produced a fire storm with rising air currents so powerful that they pulled buildings apart and sucked them upward.

On that fearful afternoon in 1910, the conditions for such a conflagration developed over an area where the fuel was measured in square miles of bone-dry timber. In Wallace and Mullan and the other settlements, in the deep canyons and on the knife-edged ridges, men were suddenly conscious of a vast silence. No leaf stirred; smoke lay flat to the earth; all sounds were muffled. It was like the hush in the eye of a hurricane. And then the hurricane struck.

Pulaski had just joined his new crew of forty men in a small clearing where they were setting up camp. Their ears were stunned by a gigantic roar. Before their eyes whole mountainsides dissolved into smoke and flame. They saw entire trees uprooted and flung at them like fiery lances. A blizzard of ashes and redhot embers fell upon them.

Many heroic deeds were done in the next desperate hours. Railroad engineers pushed burning trees off the tracks with the cow-catchers of their locomotives, drove their trains across trestles already afire, and picked up refugees all along the right of way. Iron-willed rangers forced their men to lie down in streams and pools and stay there huddled under dampened blankets until the fire passed. Men fought the fire monster step by step through the streets of Wallace while relief trains hustled their families out of the trap. There were tragic deaths and miraculous escapes.

Before his men could recover their senses enough to panic, Big Ed Pulaski had made his plan and gave his orders. They might, he thought, make it to Wallace, where there were fire engines and water and help—as good a place as any to make a stand. He must have thought, too, of his wife and their infant daughter. They might make it out of their trap if they could move fast enough to outrun the flames that were racing to surround them. Pulaski knew just how to do it—not down the canyon, in a river of superheated air that might explode at any moment, but along the ridges, where the going was rougher but also more open and there was some chance of keeping track of the fires.

He shouted against the bellowing of the fires, "Each

man grab a blanket and some grub—what you can carry in one hand. Behind me, now, in single file! Hold onto the man ahead of you. I'll get you out of this."

Alone, he could have made it easily, the expert woodsman who knew exactly where he was and just how to find the game trails that were the only pathways through most of this wilderness. Pulaski was too experienced not to know the odds. His men were already weary, smoke-blinded and choking. Most of them were flatlanders, recruited from the plains cities. Even under the spurs of terror they couldn't travel fast enough. All he had to do was force the pace and leave those who could not keep up.

"I can always take care of myself." If, at this moment, he remembered those last words to his wife he never said so.

From the ridge he peered down upon a small mountain lake in the next canyon. To the men it looked like sanctuary from the fire and relief for their parched bodies.

Big Ed knew the lake well. He had discovered it himself and named it for his daughter. "No good," he said. "Steep drop-off all the way round. No shallows. We'd drown."

By now it was obvious to all that they were not going to get to Wallace, or much farther in any direction. The fire told them that, with the bright splash of flames and its triumphant roar as it closed the last gap in the ring around them. The men huddled close to Pulaski, waiting for him to tell them what to do next. They believed in him. Not because they knew him, for most of them were strangers. Not because he wore the badge of a Forest Ranger. Scenes like this in a smoky canyon in Idaho in 1910 would be repeated many times before men would believe in a ranger simply because he was one. Pulaski held the party by sheer force of will and confidence.

He explained that there was a mine tunnel nearby. Just a prospect hole of the War Eagle Mine, long since aban-

J. S. Forest Service

Entrance to the 60-foot shaft in which Ranger Pulaski held his men

doned, but a place where they could hide out and let the fire pass over them.

How did Pulaski know there was a tunnel? No one ever thought to ask him. The answer is that it was his business to know, just as he knew the location of every trapper's cabin, homestead, trail, watercourse and pass on his district.

He not only knew it was there but found it through the muffling smoke and the rain of embers. He herded the frightened men inside and directed the building of a timber-and-blanket bulkhead to keep out heat and poisonous fumes. He took his own station at the barricade, between his men and the enemy.

For a while it was a relief to rest in the cool darkness. But then, now that there was nothing to do but wait and think, panic began to get in its deadly work. Perhaps the rotten timbers of the old tunnel would collapse and let the mountain fall in on them. Perhaps the timbers would catch fire. . . . Somewhere above the mouth of the tunnel there was an outlet, perhaps only a fissure in the rock. It acted like a chimney, sucking heat and smoke past the bulkhead. The sound of coughing was incessant. Tears ran out of smoke-burned eyes. They were all going to die in this hole, like rats, in the dark. *Run!*

For a while Pulaski held them back by force of will. After that he held them with his fists, until smoke-poisoning took the job over from him. Five men died of suffocation in that tunnel. Pulaski himself was unconscious when the fire finally passed over, the smoke and heat cleared a little and it was possible to move into the open again. The survivors led him back to Wallace and Emma Pulaski, a blind and stumbling giant.

During his long vigil at the barricade, Pulaski had taken more smoke and heat than anyone else. He never fully recovered his health or his eyesight. Nevertheless, he served another twenty years in the Forest Service, a legendary figure in his own time. While he was about it he added a word to the language of firefighters.

Big Ed was one of those rare persons who can fight the battle and remember important details. The detail he remembered was the inadequacy of men's weapons for fighting forest fire. The basic method, then as now, is to cut a trench through grass, brush, roots and litter down to bare, fireproof earth. Behind this barrier, the fire burns itself out. An axe will do the chopping but is no good for digging or scraping. A mattock or shovel is good for digging and scraping but useless for chopping. It would be much better, thought Pulaski, to have one tool that would do both.

Besides being an expert woodsman, Big Ed was also a skilled mechanic. In his own shop he fashioned a tool of tempered steel, half axe and half mattock. Today the pulaski is as much the basic weapon of the firefighter as a rifle is the basic weapon of the infantryman. How much better a monument it is than a statue that no one looks at or a plaquc that nobody reads a tool that will be used as long as there are forest fires and men to fight them!

Edward C. Pulaski was not the first man with that surname known to our history. He was a descendant of the Polish commander Casimir Pulaski, who fought on the side of freedom and became a hero of a new republic, the United States of America. So the legend of fighting men repeats itself.

Shortly after the conflagration of August 20, the fall rains did come and subdue the Great Idaho Fire. It left behind a smoldering desert never accurately measured but amounting to not less than three million acres. Half of Wallace had burned. Dozens of smaller camps and settlements were completely or partially destroyed. The known death toll was eighty-seven, but it is thought that many more than that number lost their lives—trappers, prospectors and isolated homesteaders whose bodies were never found.

Across all of Idaho and western Montana as far as Marias Pass in the Rockies, a linear distance of two hundred miles, the fire scars can still be seen. There ghost forests still lift their naked spires above a land that is green again, though not with the green towers of a

U. S. Forest Service

A 1910 photograph showing effects of hurricane and fire in the Coeur d'Alene National Forest

climax forest. The Great Idaho Fire produced a phenomenon that has never been repeated in historic times. Its gigantic smoke cloud drifted eastward on the prevailing winds, darkened cities on the Atlantic seaboard, and spread far out to sea, one-third of the way around the earth.

There have been other fires almost as devastating, before 1910 and since. In 1871 the Peshtigo Fire in Wisconsin burned less acreage but destroyed more towns and took more lives. In 1933 the Tillamook Fire in Oregon destroyed twelve billion board feet of saw timber, enough to build a million homes. But 1910 was a turning point. Never since Ed Pulaski made his epic stand have forest fires in the national forests been allowed to run completely wild. The Forest Rangers have seen to that.

2

Greatest Good to the Greatest Number

The very first national forest came into being in 1891 when President Harrison set up what was called the Yellowstone Timberland Reserve. This was a million-acre tract of wilderness surrounding what is now Yellowstone National Park. President Harrison's act was a giant step forward in conservation, but in many ways it was a step in the dark. No plans were made for doing anything with the Reserve; no one was given the job of looking after it. The land was simply set aside. Nevertheless, the creation of the reserve meant that at last, and none too soon, the leaders of our government were becoming alarmed over the destruction of our forest resources through uncontrolled fires and wasteful methods of logging. Here, then, was the first glimmer of the Forest Service idea of conservation: to protect. The other half of the idea—to use without destroying—was still unborn.

Like any other big and important concept, the Forest Service's idea of conservation was not something that suddenly appeared, full-fledged and complete. It developed over a period of years and through the work of many people. Some years and some names were, naturally, more important than others. As far back as 1866, for instance, Dr. Bernard E. Fernow was chief of a Division of Forestry that had no forests to supervise. He was a patient fact-finder who showed government leaders what our forest resources were and what was happening to them. His name is overshadowed in the history of forest conservation by the names of men like Presidents Harrison and Theodore Roosevelt and the dynamic first chief of the Forest Service, Gifford Pinchot, who built upon foundations prepared by Fernow.

Before the end of his term, for instance, President Harrison had placed thirteen million acres within the Forest Reserves. Six years went by and President Cleveland was in office before Congress passed laws that made it possible to hire men to protect this land. The first ranger of whom we have a record was an former Indian fighter and trader named Cole, who had the title of corporal-ranger. Like any other frontiersman, Cole went about his duties armed. His "district" was the Yellowstone Timberland Reserve, and it took him a whole year on horseback just to patrol its boundaries. President Cleveland had meanwhile added twenty million acres to the Reserves.

More years passed before two inspired men working together gave the cause of forest conservation its true shape and purpose. Theodore Roosevelt, hunter, soldier, explorer, author and President of the United States from 1901 to 1909, had the understanding and the authority to put to work the ideas of his friend Gifford Pinchot, a professional forester and a brilliant administrator. President Roosevelt added 148,000,000 acres to the existing national reservoir of living resources—wood, water, grass, recreation and wildlife. Pinchot put together the unique combination of men, land and ideals necessary to manage this incalculable treasure. Between them they did away with the "reserve" idea and its implication that such great tracts of land were to be locked up. In its place they created the concept of the National Forest and the practical reality of the Forest Service, of which Pinchot became the first chief.

Today we realize what an important service these two men performed for our nation. It was not so in their time. The idea of "managing" a hundred and eighty million acres of wilderness was too big and too complex for most people to comprehend. Because of Roosevelt and Pinchot we have inherited the great tradition of the Forest Rangers, from Pulaski to the newest recruit, whose mission in life is conservation. The word "conservation" has many meanings. In this book I am trying to convey an understanding of what it means to the Forest Service and what it should mean to us all. The Forest Service idea of conservation has never been better expressed than it was more than half a century ago, on February 1, 1905, when James Wilson, Secretary of Agriculture under President Theodore Roosevelt, sent Forest Service Chief Gifford Pinchot the statement of policy that still governs our management of the National Forests. The Forest Rangers, Secretary Wilson directed, should see to it that the wood, water and forage were conserved and wisely used for "the greatest good to the greatest number in the long run."

Every word in this declaration is important. Wood, one of the most versatile substances on earth, is used for everything from fuel to building material. Without water, life is impossible. Forage is food for animals and eventually for us. A plentiful supply of all three has made our nation great. They are living resources that Nature can replace every year *if we conserve them.*

U. S. Forest Service

President Theodore Roosevelt and Chief Forester Gifford Pinchot on the river steamer MISSISSIPPI, *October, 1907*

Since Secretary Wilson's day, two other resources have come in for more attention: wildlife and recreation. These may not be of such basic importance, but the creatures great and small who inhabit the wilderness are as much part of it as the trees and waterfalls. The back country would be a barren place without them. Life should be more than a grim struggle to survive. The hallmark of a mature civilization is that man should have leisure for recreation—for enjoying what he has and what he has made.

". . . The greatest good to the greatest number in the long run." Our irreplaceable living resources of wood, water, forage, recreation and wildlife belong to the nation and so to all of us. It is the task of the Forest Rangers to see that each one of us gets his fair share of the benefits now, tomorrow and a hundred years from now.

There are today approximately 183,000,000 acres within our National Forests, about one-tenth of the land surface of our country. If the National Forests were all in one block they would make an area larger than the State of Texas, an empire of forest, meadow, mountains, watercourses and life that has no equal anywhere.

It is difficult to find words strong enough to explain how important a job the Forest Rangers have, guarding and managing such a treasure. The lumber that built your home could have come from a tree that grew in a National Forest. Countless herds of sheep and cattle (which help this nation to eat better than any other) grow fat on the grass in National Forests. The headwaters of practically every important river in the United States lie within their protected boundaries so that crops may continue to be irrigated, boats to float, turbines to spin and you to drink. And when vacation time comes around you can visit the National Forests to camp, fish, hunt, hike, boat, swim, ski, climb, take pictures or simply admire the scenery.

These are our living resources, supplied by Nature with a

S. Forest Service

Young boy and black tail deer fawn, in Gifford Pinchot National Forest, Washington

lavish hand but not to be taken for granted. For they can be destroyed. The history of mankind is littered with the ruins of nations that wasted their living resources. They cut down more trees than Nature could replace. They let their livestock eat the grass faster than it could grow back. When the trees and grass were gone, rains washed away the naked soil and winds blew it away. Instead of rivers flowing gently the year round, the water courses became muddy torrents, which raged for a short time and then dried up. When their land became a desert, people moved on—or starved.

And things like this have not happened only far away and long ago to countries with unfamiliar names like Babylon. In our early days we Americans burned forests just to get rid of them. We farmed our land until it was exhausted and then moved on to new land. We killed our fish with poisonous waste from factories and cities. We slaughtered our game birds and animals. In every large city, and many not so large, the struggle to get enough pure water goes on year after year. At the same time we have destructive floods because watersheds without trees and other vegetation cannot absorb all the rain. Because of his carelessness with fire, man is still the greatest enemy of our forests.

Because our country was big and new and rich in all the living resources, our ancestors thought the supply was inexhaustible. But nothing is inexhaustible. The resources of a country are in many ways like the money in a man's pocket. No matter how much he has to begin with, if he goes on taking out more than he puts in, he will end up with an empty pocket.

But we may as well face the facts. We like that word "inexhaustible." Most people today would rather take what they want out of the pocket of living resources and let someone else worry about tomorrow. We still pollute our water, slaughter wildlife and burn forests. Conservation is something most of us have to learn. Fortunately for us, a few people are born with the ability to look at what has happened in the past and see what it means for the future, and one of these rare people was a President of the United States. Theodore Roosevelt is known as one of our great Presidents partly because he was a warrior, an outdoorsman and a statesman, but mostly because he was conservation-

Terrace Falls, Chelan National Forest, Washington

. S. Forest Service

minded. Other Presidents traveled as widely in our country as he did and were spellbound by its scenery, its wildlife, its incredible wealth in forest, grass and water. But he saw what was happening to our natural resources. He had the vision to listen to men like Pinchot and the power to make their ideas come to life. In his hands, the National Forests became just about what they are today.

Rain forest, Olympic National Forest, Washington

John Muir Wilderness Area, Sierra National Forest, California

S. Forest Service

Of course, President Roosevelt and all the early American leaders in conservation owed a great deal to foresters from overseas. As far back as 1877 Carl Schurz, an immigrant from Germany who became our first Secretary of the Interior, was urging our government to adopt scientific forestery methods. Europeans had long since outgrown the false idea of inexhaustibility. When they cut a tree they planted one to take its place, a level of conservation we have yet to attain in this country.

There are now over a hundred and fifty National Forests in the United States. The number changes from time to time as new National Forests are designated or two established Forest areas are joined. The trend nowadays toward fewer but larger Forests is possible because communications are so much better than they were in the early days. The National Forests are distributed throughout the country, but two-thirds of them are in the west, and for a very simple reason. By the time our nation was ready to accept conservation the only large blocks of public land available lay along the continental divide and the other mountain chains from the Rockies to the Sierra Nevadas.

It is no simple task to manage an area so huge and so widely scattered. As suggested by the title of its first ranger, Corporal-Ranger Cole, the Forest Service organization owes something to military models. It has a commander, the Chief of the Forest Service. His headquarters are in Washington, D.C., and he reports directly to the Secretary of Agriculture. It is a proud tradition of the Forest Service that no one of its chiefs has ever been anything but a professional forester. He is the equivalent of a general in the Army.

Under his direction are the Regional Foresters, one for each of the nine Regions, the nine main divisions of the Forest Service. Each Region is divided into National Forests whose principal officers are known as Forest Supervisors. Finally, a National Forest is divided into Districts, each one commanded by a Ranger. This is battalion level, and as any

S. Forest Service

Rough ice of Blue Glaciers (Mt. Olympus at right) in Washington

soldier will tell you, battalion level is where the fighting gets done.

Of the fourteen thousand men and women who make up the Forest Service, only about eight hundred are Forest Rangers. It is no accident that to most people the ranger *is* the Forest Service. His picture is familiar: the lone horse-

U. S. Forest Service

Aerial view of North Sawyer Glacier, North Tongass National Forest, Washington

Mt. Shuksan, Washington, 9,038 feet high, is known widely for its photogenic, rugged grandeur

U. S. Forest Service

U. S. Forest Service

Casa Diablo Hot Springs Geyser, Inyo National Forest, California

U. S. Forest Service

Bear Grass at Rock Cairns area, Clearwater National Forest, Idaho

man on a high ridge looking out over forest, canyon, lake and mountain. It is fair to state that he has the most important job in the whole organization. He is the man who ties it all together and makes it go.

But there are many jobs in the Forest Service, all of them important, and from the Chief in his headquarters in the Nation's Capital to the newest clerk in the most remote district, every member of the organization is proud to feel that he, or she, is a "forester." Strictly speaking, a forester is someone who has earned a college degree in forestry. But the word has the broader meaning too. Wherever it appears in this book I use it in this broader sense, to designate a member of the Forest Service, whatever his job may be.

A "forester" may be a white-smocked scientist peering through a microscope or watching a test tube bubble as he tries to unlock the treasures that are hidden in a piece of wood. A "forester" may be a plant detective among strange people in foreign lands, looking for trees or grasses that are resistant to disease or will thrive where other plants wither.

Herder and forest officer discuss range conditions in Payette National Forest grazing lands, Idaho

S. Forest Service

In a glass-walled lookout tower on top of some lonely peak a girl watches for that first telltale plume of smoke that means forest fire. She is a "forester," too. So is an engineer surveying a road, building a bridge, or measuring the flow of a stream. So is a wildlife specialist, planting fish in a lake; or a botanist identifying the grasses in a mountain meadow; or an electronics tech, tuning a Forest Service radio.

The smoke jumper delicately steering his parachute down onto some distant corner of the wilderness, and the avalanche buster carefully aiming his cannon through the blizzard at a target he can't even see are both "foresters." A forester may be a teacher, a detective, a pilot, a timber cruiser, a carpenter, a photographer, or a computer operator. Or a student just out of college proudly wearing for the first time the bronze shield with the pinetree emblem. The Chief of the Forest Service is the forester with the loneliest job of all. The decisions he makes have to be bold and they have to be right. For the enemies of the Forest Service are always waiting.

Unfortunately the Forest Service has human as well as natural enemies—the ignorant, the greedy and the just plain dishonest. That word "inexhaustible" is always on their lips as they try to grab for themselves some of the wealth that belongs to all of us. One of the biggest jobs the Forest Service has is education, so that people will understand just what we have in our national treasure house of timber, grass, water, wildlife and recreation. The battle is far from won, but it goes well. Under the watchful eyes of the Forest Rangers our living resources are being used, but only as much is taken in one year as Nature can replace by the next year.

Theodore Roosevelt intended that the Forest Rangers should be land managers, not caretakers. The army is small to be covering such a big territory—eight hundred rangers for a hundred and eighty million acres—but in a way matched by few organizations the Forest Service is a team. There was a time at the very first when rangers were pretty much lone wolves. Nowadays a ranger no sooner gets into trouble with a big fire or a flood or a calamity of any sort than help rolls in from every direction.

There is nothing accidental about this unique ability of the Forest Service to concentrate its forces on trouble spots. Many of the methods of communication and transportation that make it possible were pioneered or adapted by foresters of the past. And it is no accident that the rangers operate what has become a multimillion-dollar business, and do so at a profit. There are few government organizations in this or in any country that have such a record.

Most of the Forest Service's income is from the sale of timber, which in itself is a form of conservation. Forests grow old—"overmature" is the forester's word—and it is better that they should be harvested by man for his use than by fire, insects, or disease. Where the old forest stood a young one soon takes root. Thus the cycle repeats itself. The Forest Service also rents grazing rights to cattle and sheep men and leases land for such recreation developments as ski areas and vacation resorts.

Whoever you are, wherever you live in these United States, you are better off because of the Forest Rangers, who work for "the greatest good to the greatest number in the long run."

3

Forest Ranger, Modern Style

The oldtimers, Big Ed Pulaski and his fellows, carried the Forest Service through its first difficult years on their own broad shoulders, by their own woodscraft, with their own prestige. Individually, each one of them was the best woodsman in his community, just as the marshall of a frontier cowtown was the fastest gun in his. That the early Forest Rangers managed to get their jobs done with a minimum of argument and practically no gunplay is a tribute to their own reputations. Their outfit, the Forest Service itself, had no reputation; it was brand new. And the people who lived in or on the edge of these new-fangled National Forests were by no means friendly to its purposes.

Most of these people were pioneers or of pioneer stock. They had ventured into the wilderness at a time when it was every man for himself against Indians, rustlers, road agents and the forces of nature. They had the feeling that the country was theirs. They had discovered it, fought for it and survived the perils of the frontier. So some government dude was going to tell them not to dam this stream, or build that road; to graze so many livestock and no more; to cut that tree but not this tree. That would be the day!

The oldtime rangers told them the way it was, and they made it stick. They forged the image of the Forest Ranger as a man who knew his business, kept his word and backed down to no one. Since this was the case, it is something of a shock to realize that Big Ed and his kind would have trouble qualifying as rangers today.

It isn't that the modern ranger is a better *man,* but times and the Forest Service have changed. It is likely that a ranger of today would have had difficulty getting the job in Pulaski's time. The requirements are different now. The modern Forest Ranger starts young and works his way step by step up a long ladder of training and experience. He is a college graduate. In school and in the field he receives special training in silviculture, botany, geology, hydrology, ecology, snow surveying, civil engineering, and other subjects that Pulaski never heard of. He comes to his first district thoroughly tested and backed by all the experience and prestige of the foresters who went before him. It can be put in this way: an outstanding man makes his mark in any generation. If Pulaski were a forester of today he would have had all the education and special training and he might well be a Chief of the Forest Service.

Like other men, rangers come in all shapes and sizes: tall or short, thin or stocky, dark or light, talkative or quiet. One characteristic they all have in common: they are not misfits. The climb up through the ranks is hard; the demands upon a man's mind and character are great. By the time he takes on the responsibilities of a District Ranger he knows that this is what he can do and what he wants to do. Along the way he has been scanned by many pairs of knowing eyes. Like any individual, he has his strengths and his weaknesses, and they are known. Few people come to their jobs better qualified than a Forest Ranger.

From among the hundreds of rangers who man their districts from the rain forests of Alaska to the cypress jungles

U. S. Forest Service

Wildlife management in cooperation with the state: Seven hundred pounds of trout (about 2900 fish) are planted in Hopewell Lake, Carson National Forest, New Mexico

Forest Officer mapping timber stand types on aerial photos

U. S. Forest Service

U. S. Forest Service

Operator places a tree in a trench prepared by planting machine

Mary Lockett, Lookout at Spruce Mountain, Medicine Bow National Forest, Wyoming

U. S. Forest Service

of Florida, from the manzanita thickets of southern California to the pine-clad hills of northern New England, one example will have to suffice for all. His name is John Root. I choose him from among the many I have known because he stands out so boldly in my memory—because of the time and the place and the circumstances under which we met and became friends.

We first met far up the Big River in the Flathead National Forest in northern Montana, where I was exploring new country. For this sort of work, the back of a horse makes the finest of platforms. The horse takes care of following the trail, leaving the rider free to roam with his eyes and ears.

There were many things to see and hear there. The trail ahead of me was dimpled with the dainty footprints of deer and the broader, round-toed marks of elk. Below me the great wilderness river now foamed through a rocky gorge, now ambled peaceably across a park lush with grass and wildflowers. A mother grouse clucked from the thickets of huckleberry, and her fuzzball brood hastened to get out of the way of my horse's hoofs. Good mountain horse that he was, he had already halted and lowered his head as if to snort the baby grouse out of danger.

As my eyes lifted to the magnificent sweep of this wilderness I saw a sun-flash on the windows of a lookout tower. No doubt the wary sentinel perched up there had already spotted the plume of dust lifted by my cavalcade. It wasn't the right color for smoke, but a lookout takes nothing for granted. He had aimed his firefinder at it, plotted it on his map, and peered through his binoculars. At some ranger station up-country a fire dispatcher had probably already written in his log that a one-saddle-horse, one-pack-horse party was in the vicinity of Mile 15 on the mainline trail, moving upstream.

I noted that the trails on this district were unusually well kept. Roots, rocks and stumps had been grubbed out; signs were in place and newly painted. A fresh-cut log showed that a break in the corduroy across a boggy spot had been repaired. To one side, the glistening curves of a back-country telephone line jumped from tree to tree. The line was expertly strung in its sliding insulators so that no wind-whipped branch could touch it. Before ever meeting Jack Root I had formed the opinion that the ranger on this district was proud of it and knew his business.

One moment the trail ahead of me was empty except for a squirrel bouncing across in little puffs of dust. The next moment a man appeared in it as if out of the air, riding a tall horse and leading a flop-eared pack mule. I realized that he must have come out of the untracked jungle lining the trail on both sides. The bits of brush caught under the pack ropes and the scratches on the canvas pack covers proved that he had been bushwhacking.

He wore no uniform except the uniform of the wilderness, faded blue jeans, khaki shirt, scuffed leather jacket, high-heeled riding boots. He might have been anyone, a trapper looking for a new fur pocket, a guide spying out the game trails, an explorer like me. But he was a forester. That momentary glint on his left shirt pocket came from a bronze shield. Other details confirmed it. Slung under his left leg in its leather scabbard was a pulaski. Over the pommel of his saddle was draped a pair of caulked logger's boots. A trapper would have been wearing rubber soles; a guide would have had the riding boots; but only a forester, who walks as often as he rides, would have had both, especially the steel-shod loggers. This country had been touched by the 1910 fires. For a man who had to survey a telephone line or a new trail, the caulks would be just the thing for squirreling about the gigantic windfalls.

The forester waited in the trail as I approached. When the two saddle horses were head to head they snorted a challenge at each other. A good saddle horse, carrying a man, always has a feeling of importance. The pack animals, on the other hand, were bored by it all. As soon as it was plain that the two parties were going to stop and pass the time of day they began to crop the grass alongside the trail.

The forester said, "Hi."

I said, "Hi."

The forester said, "I'm Jack Root, Uncle's nephew on this ranch." To a forester, the Forest Service is always "Uncle" and a "ranch" is a district.

I gave my name. We kneed our horses together so that we could shake hands.

The ranger said, "Haven't seen you around here before, have I?"

I replied, "No. I just bought a place down on Tunnel Creek. They keep telling me that if you want to see country, the place to see it is the Big River."

"They told you right," Jack Root grinned.

He was not a tall man but very wide in chest and shoulders. There was the feeling that if one touched him it would be like touching a block of concrete. Undoubtedly his pale blue eyes had been sizing me up too. He had seen that I was well-mounted, that I packed Decker-fashion, as a mountainman should.

He continued, "Welcome to the ranch. Anything I can do to help?"

I said, "Yes, there is. I want to see the district, off the beaten track."

Root's quick, upward glance acknowledged that "off the beaten track" was the right thing to say, here where there probably wasn't another human being within twenty miles. We dismounted and went into conference mountainman-style, by squatting on the heels of our boots. Now that we were out of the saddles, our horses were relieved of responsibility. They stood companionably head to tail and swatted flies for each other. I spread my map over a fallen log. It was another point in my favor, that I had a map. I asked questions about trails and river fords, about camping spots where there would be feed for the horses, about the good fishing sections of the river and the barren sections.

Jack Root's answers proved that there was no corner of his "ranch" he hadn't visited more than once. His suggestions showed that he understood perfectly what I was up to: scouting the country for selected parties of fishermen and hunters.

He made a pencil mark on the map. "You might turn off the main line here and go up Cy Creek. Quite a game pocket in there, but few guides will tackle it because it's a rough climb on the other side. You can make it with the kind of stock you have, and you won't be bothered by company. The fire trail peters out at the head of the creek, but you can bushwhack on over the divide and hook onto this trail here, come out through Three Forks." His pencil had traveled five inches over the map which, on the ground, translated into three thousand feet up, two thousand feet down and five miles of the roughest country under the sky.

Between us we laid out a circle tour of the wilderness that would touch all the best hunting and fishing spots and take a week. He was under no obligation to go to all this trouble for a stranger. He was doing it because he loved this wilderness and, since I'd shown a real interest, because he wanted to share its glory with me.

"I'll tell the dispatcher where to expect your campfires," he said as I folded up the map. Otherwise a lookout would aim his firefinder at the smoke, and five minutes later a smokechaser would be hot-footing it up the trail. "By the way, are you available for a fire call?"

It was a trap question. If I was what I appeared to be I'd know that I was automatically available for a fire call. I fished the Fire Qualification Card out of my shirt pocket and handed it over.

"Hmmm. Smokechaser, scout, crew boss, packer. Six head of horses. You'll be handy to have around."

Before we parted I agreed to put up a heavy smoke from

a spot not on my campground list. Root explained that he had a green lookout covering that area and wanted to check up on him. Plainly this ranger thought all the time, generally one jump ahead.

For several years—a long time as such things go in the Forest Service, where transfers are frequent—I knew Jack Root. We became friends. Close to the Forest Service but not a part of it at that time, I had an unusual opportunity to watch a modern Forest Ranger in action. By this time, the 1940s, the Forest Service was a strong and respected organization. It had gained prestige through achievement:

Ranger LeRoy Sprague using increment borer in timber stand on side of West Mountain, Boise National Forest, Idaho
U. S. Forest Service

by fighting fire and winning, by preventing floods, by stopping the waste of timber, by persuading grass to grow again on ranges that had been overgrazed to the point of destruction, by building roads and trails into the wilderness so that all men could actually see and enjoy their heritage. By Jack Root's time a Forest Ranger was someone to be respected simply because he was one.

A ranger of the 60s is not much different. He is likely to

Participating in a Forest Service totem preservation project, CCC enrollee Gebrt Denny, Sr., and his grandchildren pose at base of the Giant Oyster Totem, Tongass National Forest, Alaska

S. Forest Service

be a few years younger than the rangers of Root's generation, because the training is better as a result of the experience of Root's generation. The job grows more complex as more people use the Forests in more ways. A District Ranger's office nowadays usually has more assistants in it than a Forest Supervisor's had twenty-five years ago. But the most noticeable changes are in the hardware. The technological advances in forestry are described in later chapters, but a few can be mentioned here as samples. Where Root made do with the cranky tube-type radios of the time, the ultramodern ranger has the transistorized marvels of today. No inventor has been able to replace the pulaski, but it has been reinforced by a host of other hand tools for special situations. Perhaps the greatest innovation in tools is the power-driven chain saw, which can gnaw down a tree in a fraction of the time it takes with the crosscut saw of song and story.

On the day when he welcomed a stranger to the Big River country, Jack Root was absolute ruler of a mountain empire a hundred miles long by fifty miles wide. Within the boundaries of this District, the size of the State of Connecticut, there was not a single mile of highway, not a post-office, not a store, not a town. But his headquarters at the fork of Big River was the hub of a thousand-mile network of trails. He had two airfields—or at least what the bush pilots of the Forest Service call airfields. His cunningly placed lookout towers, smokechaser cabins and work camps were tied together by radio and telephone. From his bulging warehouses at Three Forks he could supply a small army of firefighters. This wilderness empire was his responsibility, to conserve and defend. If anything happened to it, he would answer.

Actually, the wilderness had only two enemies capable of doing it much harm: man and fire. Jack Root was well prepared to deal with both. Through all that part of Montana the game hogs, the fur poachers, and the timber and

S. Forest Service

Helicopter pilot receives instructions from the fire boss before takeoff from the Chilao Heliport

grass thieves had learned to steer clear of the Big River District. An old lobo was on guard there who knew the country and all the tricks better than they did.

As for fire, it was Root's proudest boast that he could get a man to eighty percent of his district in less than an hour, to any part of it in less than four hours. He understood not only the methods of firefighting but also the men who did the dirty, dangerous and exhausting work. I remember coming into his headquarters late one afternoon when a lightning storm had touched off smokes all over his district. There were a dozen separate fires on Spruce Mountain alone.

Since dawn I had been guiding one crew after another through the jungle of devil's club and alder to where some lightning-struck tree flew its banner of smoke. At the door

ANGELES
CAL-ROTORS
INC.

to the office I met another smokechaser who had been doing the same thing. All I remember about him is that his name was Red and that he was hollow-eyed and shaking with fatigue, walking jarringly on his heels as a man does only in the last stages of exhaustion. For that matter, I was not much better off.

For no apparent reason, Red suddenly went to his knees. He tried to get up and couldn't. Two things were obvious. Red was finished and nothing would make him admit it. I stood him on his feet and we went into the office together. Root turned from the fire map studded with red-headed pins and took a brief look at us.

"Monty," he said, "You take a crew to this one." He pointed to one of the red pins and gave me the map coordinates. "Red, we're pretty well caught up. You hit the sack."

In that single glance, Root knew that I had one more fire-run in my legs and that Red had none. But how? It is one of the mysterious elements in that mysterious quality known as leadership.

Always a seeker after new ideas and better methods, Jack Root had a hand calloused in the process of ushering the Forest Service into the age of airpower. As he told the story, he had been a ranger with the factory shine still on his badge back in the 1920s, the first really bad fire years in that area since 1910. Already the Forest Service was using aircraft to scout fires. Some daring thinkers were even talking about dropping equipment and supplies to isolated crews, but not very seriously.

Down on the St. Joe, Root was fighting a fire in a 1910 burn. In the years since that disastrous summer the once green and mighty forest had turned into a legion of ghost trees, a hundred feet tall and three feet thick at the butt.

U. S. Forest Service

A heliattack crewman ready to take off. He will jump from the helicopter hovering 5 to 10 feet above the ground.

One by one, as their root systems decayed, the silvery trunks had toppled over, this way, that way, every which way. In places these monstrous jackstraws were piled twenty feet deep. Brush and weeds and second-growth trees had pushed up through the pile. Now the whole mess was on fire.

"The fact is," said Root, "we were trapped. If the fire got away we'd had it. The thing we needed most was saws. Those old tamaracks had seasoned hard as iron. It took an hour to cut one with an axe. So there was this scout plane buzzing around over us, not doing anybody any good that I could see. So I asked him how about loading up a few bundles of crosscuts and dropping them to us."

Incidentally, the Forest Service pioneered the portable, battery-powered radio telephone. In those days it bore little resemblance to the modern transistorized handie-talkie that one hangs on his belt or from a neck cord. Jack Root's "portable" was a heavy load for two men. The batteries didn't last long, the tubes were continually blowing out, and its range would make a modern radioman laugh. But one of those early-day portables brought Jack Root his saws in the first air-drop of cargo in firefighting history.

Inexpertly packed, the bundles broke apart in midair. A rain of steel ribbons, lance-toothed and six feet long, fell down on the desperate firefighters.

"I was never so scared in my life," chuckled Root. "We crawled under logs, into the brush—dodged like rabbits. Some of the boys got carved up a little, but no one had his head cut off. After a while we rounded up the saws and put the fire out."

Like Big Ed Pulaski, Jack Root was skilled with his hands. He had invented a trail-grading plow that had a lot to do with the excellent maintenance I had noticed the moment I came onto his district. He also inaugurated a system, which eventually became Service-wide, of measuring and marking his trails mile by mile. Not that old guess-

U. S. Forest Service

The Forest Service's timber business is its best means of self-support. Forester Kenneth Estes measures the merchantable height of a Ponderosa pine to be marked for cutting in Trinity National Forest, California, and then uses a diameter tape (below) to measure its diameter.

U. S. Forest Service

work or scaling on a map that may have led to the saying that Forest Service miles are lo-o-o-o-ng miles. Root's measuring was done by a mule-drawn pedometer, a bicycle wheel, so many revolutions to a mile. This system was one of the reasons Root could get a man to any place in eighty percent of his district in less than an hour. It was the old time-and-distance formula, the solution in this case being the fact that a man can travel four or five times faster on a trail than he can cross-country, especially Big River country.

When a fire broke out, a three-man team went into action: the smokechaser nearest the fire, the dispatcher at District Headquarters and the lookout who spotted the fire, all tied together by radio or telephone. The lookout reported to the dispatcher the direction—azimuth—from his tower to the fire and his estimate of the distance. The dispatcher asked other lookouts for a cross-bearing. On his big map and plotting board he planned the smoke-chaser's route, not the shortest but the *fastest.* While this was going on, the smokechaser kicked his heels impatiently and stared at that legend stenciled on his fire pack that reads, "Minutes Count, Let's Go."

He didn't kick them long. A skilled fire dispatcher makes short work of reducing the distances and directions from the lookout towers to distances and directions from a smokechaser cabin. His order to the firefighter was terse: "The fire is in the northeast quarter of Section 23. Take Trail 1 to Trail 9. At Mile 2 on Trail 9, take off on azimuth 188. Good luck."

Unlike Pulaski, Jack Root never became famous. Superb forester and organizer though he was, by his time he was only one among many. To my mind he was outstanding. I often wondered why he didn't go higher. The jobs of Forest Supervisor, Division Chief, even Regional Forester, were within his capabilities. Over a campfire one night he told me why. He was that rare being, a contented man. He wanted no empire other than his wilderness district.

4

The Hammerheads

The time-and-distance problem exemplified by Forest ranger Jack Root's adventure with the saws is one with which the Forest Service has wrestled in various ways from Pulaski's day to this.

A ranger may be going out to mark a timber sale, a job for which he can carry all the equipment he needs in his pockets. Or he may be moving a thousand-man army into battle against a fire with a supply train that includes everything from bulldozers to sleeping bags. No matter what he is planning to do, he has to solve the equation of time-and-distance. The answer to that equation is more urgent for a ranger than it is to other men, for in wilderness distances are always great and time is always short. There is nothing he can do about the distance itself; a mile is five thousand two hundred and eighty feet cross-country, by trail, on a road or through the air. So, in its simplest terms, the problem is this: how long does it take to move a man or a pound of freight a mile, and how can that time be reduced?

In Pulaski's day, roads and trails were few and widely scattered. Bushwhacking in rough country and dense timber, a man's rate of march could be painfully slow, as little as half a mile per hour. It wasn't fast enough, the rangers dis-

U. S. Forest Service

The old days: A crew of Foresters in camp, 1907

Below: Forest Service Ranger Station, Aquachiquita Canyon, 1908

U. S. Forest Service

covered—not fast enough to catch up with a fire; not even fast enough to get out of the way of one. In 1910 many people lost their lives simply because the fire outran them. It even outran the machines of that day. An engineer of the Great Northern Railroad has told of starting the long climb from Belton at the western edge of Glacier National Park to Marias Pass at the same time as the fire. The fire, riding the wind in mile-long strides, beat his train to the summit.

The rangers realized immediately that transportation, just the getting from one place to another, was almost their first order of business. In the years following 1910 they built hundreds of miles of road and thousands of miles of trail. Some of their work was crude by modern standards. Brush and windfalls were cleared out of game trails that had first been pounded out by deer and elk. A road was scratched across a township with pick, shovel and scraper. But even this improvised transportation system greatly improved the mobility of the forest army. Over any kind of a trail an experienced man on foot or on horseback could average four miles per hour. Even on the rutted, bumpy roads of the back country he could do twenty miles per hour by car. In an age of jet aircraft and superhighways such speeds may not be very impressive. But to a smokechaser hot-footing it to a lightning strike, the difference between half a mile per hour and four miles per hour is conclusive. No matter how it started, whether from a lightning bolt or a camper's match, every fire begins small. If he can get to it in one hour instead of in five or six hours, the man with the pulaski can put out that fire before it gets beyond the smoldering stage.

The year 1929 was as bad a fire year as 1910. Yet in the same area where Pulaski fought in vain, the forest rangers used their improved road and trail system to reach a thousand lightning-set fires. They cut the loss from three million acres burned in 1910 to five hundred thousand burned in

1929. Though it represented a vast improvement, this record did not satisfy the foresters.

In particular, it did not satisfy the newly appointed Regional Forester, Major Evan Kelley. When the fall rains finally came to relieve the exhausted firefighters in that near-disastrous year he called for reports from the leaders in the field. They told a story that he had already begun to

The old days: Fighting forest fire on the Wallowa National Forest, Oregon, 1908

U. S. Forest Service

suspect from his own investigations. The Forest Rangers now had the network of roads and trails they needed, the equipment, the leadership and the manpower. One important thing they still lacked was a freight-moving system equal to the emergency. They had enough saddle horses and pack animals to do their ordinary work. But in a fire year, when the rangers were forced to recruit every half-trained animal and amateur packer in the region, the supply system broke down. Men got to the fires, but the equipment and supplies they had to have in order to fight were strewn all over the countryside. Regional Forester Kelley made his rangers a promise: A more efficient method of moving freight in the back country would be found, and quickly.

It was a fortunate choice that brought Kelley, an ex-cavalryman and commercial packer, to Region One of the Forest Service at this crucial time. He was better qualified by training and experience than most men to spot the flaw in the firefighting organization and find the remedy. Kelley's solution was in two parts, the first of which was an invention. Several years previously, a commercial packer named Decker had devised a new kind of pack saddle.

Since the dawn of history, men have been in one manner or another tying their loads to the backs of animals. Without that method of carrying equipment and supplies and so increasing their range of operations, men could not have won this continent. But the ridge-topped, round-sided, jiggling back of a horse or any other beast of burden makes an awkward platform. The load slips, rubs, shakes apart, overbalances, or gets scraped off. Before the Decker saddle, the most efficient method of packing was the sawbuck saddle and diamond hitch, borrowed from those master horsemen the Plains Indians, which had been improved upon and universally used on the frontier.

As a method of lashing cargo to the back of a horse or a mule, the Decker saddle and the hitches that go with it were revolutionary. Regional Forester Kelley, formerly a commercial packer himself, instantly recognized the Decker as a breakthrough. For the first time in history, weight became the only limitation on what could be carried on the back of a pack animal.

The most awkward loads for packing are long, thin ones, like a piece of pipe twenty feet long, and the small and heavy ones. Such a load was the three-hundred-pound cookstove that the famous pack mule, Pawnee Bill, lugged the seventy-five jiggling miles from Coram, at the mouth of the South Fork of the Flathead River, to Big Prairie, in the heart of the Marshall Wilderness. With the Decker method, the size and shape of the load made little difference. Whatever the load was, it could be tied on and would stay there.

It is difficult to explain to anyone unfamiliar with packing why the Decker method is so much better than the old, reliable sawbuck saddle and diamond hitch. I had it explained to me in the most convincing way, by demonstration. A diamond hitch man myself, I happened to be loading up beside a Decker man one day. To begin with, I liked the look of his saddles. The dinky sawbuck perches high on a horse's withers like some experimenter's first try at designing a saddle. The Decker sat low and solid and had a canvas hood stuffed with grass to protect the horse's ribs.

The sawbuck load is in three parts: the two side packs and the top pack, which sits high in the middle. Over the whole goes the diamond hitch, a complicated pattern of rope that is not only difficult to tie correctly but also time-consuming to put on or take off. This top-heavy load has to be balanced exactly in weight and leverage or it will capsize, an event guaranteed to turn the tamest pack horse into a bucking bronco.

The Decker man tied his load into two canvas-wrapped bundles. Then he slung them low on the sides of his horse in the cargo hitches. The cargo hitch is a beautifully simple knot that grows tighter the more you jiggle or pull on it but releases at a flick of the packer's wrist. I noticed that the loads didn't balance exactly, for the saddle was canted, and I wondered what the Decker man would do about that. It would be a tedious business to unwrap and repack everything. What he did about it was to cinch the heavy load a little higher than the light load. Leverage took care of the weight difference.

I was more than convinced. I never threw another diamond hitch as long as I was a packer. To his outspoken annoyance—Major Evan Kelley had a notable temper—the Regional Forester found that some of his old-time packers were not so easy to convince. A generation brought up on sawbucks and diamonds was not about to adopt this newfangled system on the word of some dude from headquarters. They revolted.

Always a man for direct action, Kelley brusquely summoned the mutineers. He told them to choose their best man. Between this man and a Decker packer he staged a

Method of packing two or more iron cots on a Decker packsaddle

U. S. Forest Service

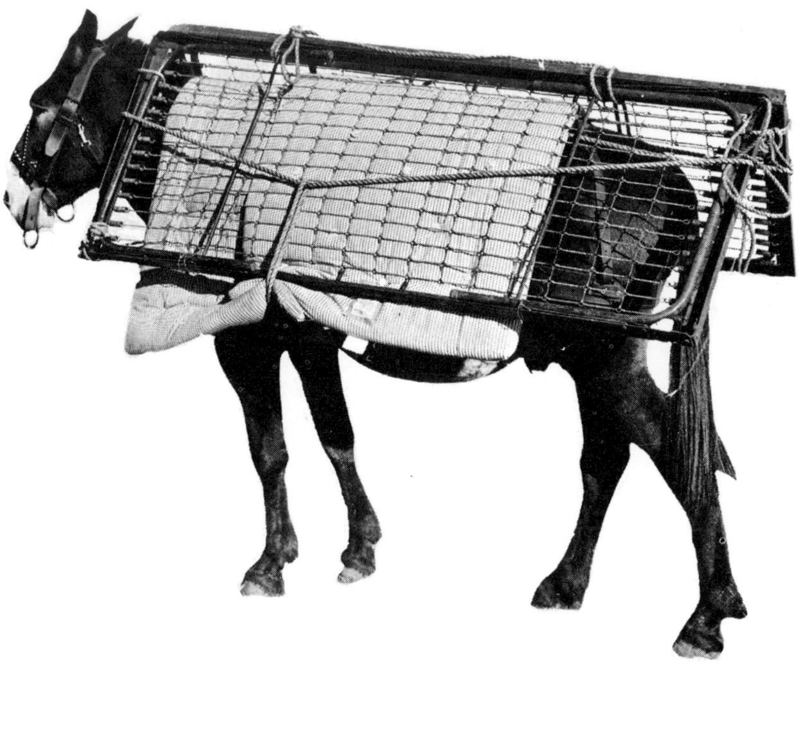

competition. Each had the same number of animals and the same cargos. The Decker man had his string loaded and ready for the trail before the sawbuck man had thrown his first diamond. That was the end of the revolt.

The second step toward the solution to moving freight in the back country was finding a better type of pack animal. The usual pack string in 1929 was a motley collection of broken-down work horses and half-wild mustangs rented from some guide or rancher. Kelley attacked this problem in his usual bold fashion. As an ex-cavalryman he knew that a mule—a bad-tempered, flop-eared, incredibly sure-footed and hardy cross between a donkey and a horse—makes a better pack animal than a horse. Near his headquarters in Missoula, Montana, he established a Remount Depot where the Forest Service bought, trained and eventually raised its own mules. In time the Remount developed a supermule, nicknamed (not always affectionately) the "hammerhead." To a mountain man the hammerhead pack string was a thrilling sight. All the same color and size, carrying enormous loads, the mules could shuffle along the high trails at a speed that made his eyes pop.

The hammerheads could carry a three-hundred-pound load, twice the capacity of any other pack animal, day after day over the rough trails at a steady four miles per hour. They made all previous methods of back country transportation obsolete.

They moved so fast, in fact, that no ordinary horse could keep up with them. This was serious. No man wanted to ride a mule, with its rough gait and razor back. More important, that hybrid creature is peculiar. A mule will follow a horse wherever it leads, but not another mule. So Kelley imported stallions of a little-known Hungarian cavalry breed and developed the mountain horse—short-backed, long-legged and broad-chested. These animals were intelligent and easy-gaited. And they could outwalk anything on four legs, hammerheads included.

One problem always led to another. A saddle horse or pack animal traveling the rocky trails under heavy loads must wear shoes. The hammerheads traveled so fast and so far that they wore out a set of shoes in six weeks. Shoeing three or four a day, at the normal pace, the blacksmiths couldn't keep up with the demand.

Horseshoeing is an art as old as packing. In fact, one probably begot the other. It has its own traditions and methods. There was no way to speed it up. So said the craftsmen. Kelley thought otherwise. Presently he had blacksmiths who could shoe twenty head a day—much to their own amazement, probably.

Among this new breed of blacksmith the admitted champion on the Flathead was Dan Hotel. I happened to meet him one day at Black Bear Ranger Station on the South Fork. Having heard tales of his skill, I asked if he would let me watch him work. Putting on a set of shoes is a touchy business for both man and animal and few blacksmiths will tolerate an audience. Dan didn't object. It was my good fortune to see a feat of strength and skill that would be difficult to top.

There was no stock at Black Bear in need of shoes except Dan's own pack mule, Aura. There were tales about her, too. This hammerhead was a giant of her kind, standing seventeen hands tall, which is almost six feet, and weighing over a thousand pounds. She was such a troublemaker that even the Forest Service linepackers had given up on her. Dan Hotel took her over because she could carry his tools, a two-hundred-pound anvil on one side and his tool-box, equally heavy, on the other. Since the blacksmith stood not over five feet four inches and was almost as wide as he was tall, every ounce of it muscle, to see them together was to think instantly of David and Goliath.

All business, Dan spread out his equipment, tied his mule to the hitchrail and went to work. This is the proper approach. A horse likes to be talked to and fussed over, but

U. S. Forest Service

Lunch at Meadow Creek

Bob Marshall Wilderness Area

U. S. Forest Service

not a mule. At first one would have thought Aura the most docile creature on earth. Ears drooping, eyes half closed, she stood there quietly while Dan picked up a forefoot, pulled off the old shoe, trimmed the hoof with a few expert strokes of the rasp, fitted the new shoe, and drove in the nails. But I thought there was a calculating look in those half-closed eyes.

A horseshoe nail is very sharp and beveled so that when driven into the horny, insensitive part of the hoof it turns and comes out along the rim. Aura proceeded to demonstrate what can be done with such a weapon. The moment Dan reversed his hammer to twist off the points with the claws, she jerked her foot away from him. The nails went through his leather apron and into his leg.

The blacksmith let out a bellow of rage and pain. At the same time he swung round, got his shoulders under Aura's belly and heaved. Eyes wild, ears and tail flopping, that enormous mule rose into the air, turned over and landed on her back with a thud that shook the ground. In a flash Dan had her legs lashed to the hitch rail. And in that upside-down position he finished shoeing her. It is said that from that day forward even a cook's flunky could shoe Aura the hammerhead.

Shoeing mules four times as fast as it had ever been done before was a relatively simple problem for Evan Kelley. Distance was a greater challenge. Region One of the Forest Service extended from eastern Washington through Idaho into Montana. No one could foretell where fire would strike. Pack strings might be needed hundreds of miles from the Remount Depot. Without a means of bridging that gap Kelley's magnificent transportation system was only half effective. He presented his engineers with the job of designing a gondola-type truck that could move a complete pack train, nine hammerheads and the packer's horse at speeds up to sixty miles per hour.

U. S. Forest Service

Forest Service reconnaissance party in Stillwater Canyon, 1938

Pack train in the Selway-Bitterroot Wilderness

U. S. Forest Service

Picture a forester with a going fire on his hands. The crews are on their way in, carrying a few tools. But this is no little spot fire that will lie down the first time a man swats it with a pulaski. It is moving, getting up a head of steam. Equipment will be needed for a long fight: pumps, power saws, bedding, food, tents, radios, more men, more tools for those men. The call has gone out and the forester knows that help is on the way, but *when will it arrive*?

At the Remount Depot, it is midnight, and the warm summer darkness lies over the sleeping camp like a satin blanket. There is no sound except for the stamp and squeal from the corrals. There the pack strings on call are lined up at the feed racks. A single lamp glows in the office, where the fire dispatcher drowses over a magazine with headphones clamped to his ears.

He jumps as the headphones crackle. He presses a button, and an alarm bell startles the night. Floodlamps rip the darkness apart, and men pour out of the bunkhouse. In the garage a giant engine clears its throat; then the gondola drifts across the yard to the loading chute. The tailgate thuds down. Back in the corrals the pack string on deck is already in the haltering alley. The packer leads his saddle horse aboard, and the hammerheads come crowding after, grotesque in their dust goggles. The dispatcher runs out of the office with a slip of paper in his hand.

From far down the smoky valley the waiting forester hears the first drumming echoes of a big engine with the throttle wide open. The echoes swell and fade, the way sound does in the mountains, but always they come closer. Then one of the most monstrous things ever wheeled on rubber comes around a bend in the highway. The gondola motor howls as the driver double-clutches into a lower gear. The truck goes by in one long flash of green paint surmounted by the goggled faces and flopping ears of hammerheads standing three abreast.

The forester catches a glimpse of the legend stenciled on the side of the machine:

U.S. FOREST SERVICE
REMOUNT PACK TRAIN

At such a moment, a forester can't help choking up. I know from personal experience.

By 1940, Evan Kelley's Remount Depot had two thousand hammerheads on the trails of Region One alone. With them and the Decker pack saddle, the Forest Rangers were moving equipment and supplies to their back-country armies at a rate never approached anywhere at any time. A hammerhead pack train, the Forest Service's streamlined version of the most ancient transportation system on earth, could move a ton and a half of supplies and equipment at a steady four miles per hour.

The Forest Service packers, who could load a full string in less time than an ordinary packer could put a diamond hitch on one horse, became the aristocrats of the high trails. Like horsemen everywhere, they acquired colorful names—Big River Dollinger, Yellowjacket Paulen and Red Mountain Jones. The stories of how they got those names are even more colorful.

Jones got his for his exploit in taking supplies to the Red Mountain fire through trailless country everyone else thought was impassable to pack animals. Yellowjacket Paulen was peacefully leading his string along a back country trail when his horse brushed against a hornet's nest. The angry yellowjackets swarmed to the attack. The situation had all the markings of the worst wreck in the history of packing. Paulen did the only thing he could: he led his string out of the danger spot at full gallop. With the lead ropes stretched tight there wasn't anything the frantic mules could do but run. Paulen saved his pack train, but it took him two days to gather up the equipment scattered along three miles of trail.

S. Forest Service

Planting 25,000 rainbow fry at Lagunitas, Carson National Forest, New Mexico

Dollinger was crossing the turbulent Big River. One of his mules lost its footing in midcurrent, and instantly the whole string was swept off the ford. Dollinger, no swimmer, could have saved himself by simply dropping the lead rope, for he was close to shore. Instead, he elected to fight. He went down the river with the struggling mass of animals. The fully loaded mules were top-heavy and in danger every moment of turning upside down. Somehow Dollinger floundered from one to the next, cutting the pack ropes with his knife. How he avoided getting kicked to pieces is one of those mysteries no one can explain, not even Dollinger himself. But he saved every one of his hammerheads.

By 1940, Ranger Jack Root could reach any part of his Big River District in less than four hours. In less than a day he could muster a hundred men at any trouble spot, and in two days he could deploy a thousand fully equipped. All this in a wilderness which twenty-five years earlier had been almost impenetrable to anyone but a man on foot, carrying his equipment on his back.

As it happened, 1940 was another year of reckoning for the Forest Service. Again the hot winds blew, and again the thunderheads marched across river, mountain and forest. Lightning set three thousand fires in Region One, and the Forest Rangers pounced upon every one of them and put them out. On the backs of hammerheads, five million pounds of food alone went to the firefighters. The red enemy's toll went down to one-eighth of the 1910 losses.

The day of the great pack trains has waned. In the same year that the hammerheads broke the back of the worst fire year on record, airpower was coming into its own. But the deeds of the hammerheads are still told around the campfires of the Forest Service.

5
Airpower

The quarter-century from 1915 to 1940 was an exciting era marked by one scientific breakthrough after another. All of them are taken very much for granted today: mass production of automobiles and the bulldozers that built the roads for automobiles to travel on; the long-distance telephone; broadcast radio; high-tension electric power; talking pictures; transport aircraft. At no other time in history until the age of rockets has man made such great advances in solving the ageless problems of time and distance.

Many of these technical marvels were made to order for the Forest Rangers, who eagerly adopted them. The young and lusty Forest Service did not accept the idea that a bureaucracy is supposed to be ultraconservative and ten years behind the times. The foresters even added refinements of their own to the new developments. From the Army Signal Corps they borrowed the one-wire telegraph-telephone system that uses the earth as one side of the circuit. It was inclined to be noisy, but it cut in half the amount of wire needed, an important feature where all the material had to be lugged by pack animal. The engineers who developed broadcast radio did not believe that a radio telephone could be powered by batteries and made small enough to carry into the wilderness on the back of a man or a hammerhead. But the Forest Rangers needed this mobile means of communication so desperately that they made it work anyway. Thus they had a part in the development of portable communications systems whose end-products today are those transistorized marvels that broadcast from space.

Among these scientific and technical advances, none has had so spectacular a history as flight, man's dream since immemorial times. With jetliners filling the skies and spacecraft orbiting above, it is difficult to comprehend that the first successful powered flight took place in 1903, and that Charles Lindbergh flew from New York to Paris in 1927. One lifetime spans the whole adventure.

No one, of course, could think more longingly of aerial transportation than a forester. It would be the solution to

Pilot Jack Alkinson (rear seat) and Ranger Charles W. Clover, observer, starting on test trip over San Isabel National Forest, Colorado, August 28, 1920

U. S. Forest Service

. S. Forest Service

The gondola of a Pony Blimp, ready to take off on a fire patrol over the forests of Southern California. Angeles National Forest, 1921

so many of his heartbreaking problems of time and distance if a man could leap over the pathless jungles, unfordable rivers and cliff-walled canyons to his goal! Even if he could do no more than cruise overhead and see exactly where a fire was, how big it was and what it was doing, that would be a dream come true.

Aircraft passed from the gadget stage into the practical under the stimulus of World War I. After the war, the forester's dream did come true—or, to put it more accurately, foresters made it come true.

It was nothing that came to pass overnight. By 1925 aerial fire patrols were roving the western skies from California to Idaho, a joint effort of the Forest Service and the Army Air Force. Scouting was the first—and for a time the only—use of aircraft by the Forest Service. These patrols spotted fires, mapped timber and searched for lost persons. The ambitions of the foresters certainly did not end there. The story has already been told of how Jack Root engineered what may have been the first aerial cargo drop in history. Others had the same idea. But bombing the wilderness with cargo was not good enough. A bundle of saws could survive such rough treatment. But there are many other things a firefighter needs, from a motorized pump to a crate of eggs. What the foresters yearned for was a system for gently lowering any sort of cargo—even, some of the more imaginative were thinking, a man—out of a speeding plane onto a rough mountainside.

The solution was to marry one invention to another: aircraft to parachute. The parachute was hardly a new invention. That Renaissance genius Leonardo da Vinci understood the principle—the big canopy that rides a man down on a cushion of air. As early in the modern era as 1916 a man named Adams took out patents. But like the plane, the parachute did not really come into its own until after World War I. In the 1920s no county fair was complete without a troupe of barnstormers who did acrobatics on the rickety planes of that day and topped the performance off with a parachute jump.

The foresters took it from there. In 1929, spurred on by a bad fire season, they made the first parachute drops of cargo. By the early 1930s the technique was so well established that a fire boss in trouble far out in the back country needed only to speak into his radio or telephone. In a matter of minutes the chutes would be drifting down with anything from a complete camp for a hundred men to a bundle of pulaskis. It was time for the next step forward.

U. S. Forest Service

Two smoke jumpers pack cargo chutes

Preparing cargo for air drop on Basin Rim fire

S. Forest Service

T. V. Pearson, of the Intermountain Region in Utah, was one of the imaginative thinkers. In 1934 he set up a parachute-jumping demonstration. But there was opposition from officials who felt that this was a harebrained scheme indeed, a belief that the daredevil antics and frequent crashes of the barnstormers did nothing to correct. The idea of transporting firefighters by air and parachute went back into mothballs.

The next year, however, the Washington Office of the Forest Service organized an Aerial Fire Control Project. The goal was aerial bombing of fires with water or chemicals. Experiments were at first carried out in California. In 1939, the project moved to the North Pacific Region and a fortunate encounter with David P. Godwin, Assistant Chief of Fire Control. He was another of the progressive thinkers. By this time it was obvious that the aerial bombing scheme was a failure. The aircraft of that day simply didn't have the performance. On Godwin's recommendation the project funds were put into a parachute-jumping experiment.

Godwin was able to go far beyond Pearson's demonstration. He contracted with the Eagle Parachute Company of Lancaster, Pennsylvania, for parachutes, riggers, professional jumpers and instructors. One of the first jobs was designing an armored jumpsuit, a must for men who were as likely to land in the crown of a tree or on a boulder-strewn hillside as on level ground. The leader of this group was Frank M. Derry, who was to earn a special place for himself in smoke-jumping history by inventing the slotted parachute. By manipulating the slots in his canopy a smoke jumper can steer himself around the sky almost like a glider.

By the end of the summer of 1939 the test group had proved that a parachutist could land anywhere in the forest. The experiment was such a success that for 1940 the Forest Service decided to set up a regular training base. The site chosen was Ninemile, a former Civilian Conservation Corps camp near Missoula, Montana. The choice was logical, for

Missoula was still, as in 1910, the hub of the great wilderness areas where smoke jumpers would be most valuable. Word went round among the husky young smokechasers of that season that volunteers were wanted for the smoke-jumper school. They flocked to sign up, but many were disappointed because the doctors who gave the physical examinations were choosier than a football coach.

Most of the trainees were College of Forestry students getting practical experience to go with formal education. They were young, tough, eager and in top physical condition from bushwhacking through the mountains. What happened to them when they reported to Ninemile was probably the last thing they expected. The first thing they did was go through a mental and physical conditioning program that would have daunted a professional boxer. Along with the fiendish set of calisthenics devised by Frank Derry and the other instructors, there was re-education. All of their previous training in map reading and use of the compass, for instance, had been on the ground. Now they had to learn airmanship: aerial map reading, how to gauge air currents, first aid and survival technique, how to pack a parachute. Beyond that they practiced endlessly on apparatus ingeniously designed to let a man make a parachute jump without leaving the ground. These devices were, incidentally, the forerunners of a whole host of training aids in use nowadays to simulate reality and give training in any technique without incurring the expense and danger.

One of the training aids was a sort of derrick. The student smoke jumper hooked his parachute harness to its dangling rope. Then he jumped from a platform, and after he had fallen the correct number of feet the rope stopped him with the same violent jerk as an opening parachute. He learned that he must jump straight and clean or get a jolt that rattled every bone in his body.

A parachute landing is the same as jumping off a twelve-foot wall. The smoke jumpers practiced for it by jumping off a structure like a diver's tower. But instead of water, their landing was on hard ground. They learned to cushion the shock by coming down relaxed and rolling.

In the wilderness, the top of a tall tree would, on occasion, be the smoke jumper's landing place. To prepare them for it, the instructors hung the students from a clothesline arrangement, a dozen at a time, like so much laundry. While dangling in midair, the smoke jumper fished an escape rope out of the leg pocket of his jumpsuit, threaded it in a precise manner through the straps of his parachute harness, and attached it to the shroud lines. Then he pressed the quick-release buckle of his harness. If all went well, the harness came free and he could slide to earth on the escape rope. But if he made a mistake he wound up hanging ignominiously head down in a tangle of webbing and rope. His reward was the raucous comments of his mates. Like brave men everywhere they made a game of something that was deadly serious. That same mistake, made on a fire jump in the wilderness, could cost a man his life.

When even their perfectionist instructors were satisfied, the student smoke jumpers began making actual descents. At first the landings were on the level airstrip. In response to orders from the ground by loudspeaker, they maneuvered the steering slots of their parachutes and came in with a thud and a grunt, to roll and spill the wind from their canopies. They found that going off the jump tower twelve feet above ground was not quite the same thing as stepping out of a plane onto three thousand feet of thin air. Not all of them could accept that difference. If a man refused his first jump he could signal to go round again. But if he froze a second time he was finished as a smoke jumper.

I remember the remark of one high Forest Service official while we were watching a class of smoke jumpers make their graduation leap. The plane approached the field in a shallow dive, then leveled off. We could see the first jumper poised in the cabin door, inhuman in his mask and armored

J. S. Forest Service

Smoke jumper

Heliattack fire crewman jumps from a Sikorsky Helicopter in the San Gabriel Mountains, California. He will lie prone until the helicopter passes by.

J. S. Forest Service

jumpsuit, with parachute pack fore and aft. We could sense the rest of the team, standing single file behind him, in the shadows of the cabin. The jumpmaster signaled, and parachutes began blooming against the big sky of Montana as evenly spaced as beads on a string.

"Pretty fair jump," said the ground controller.

"They're heroes," said the official, "every one of them."

The most frequent question asked at Ninemile was "When do we jump on a fire?"

The only answer the instructors could give was a shrug and a wise look, for they knew no more about that than the trainees. Still moving cautiously into this new medium, the Forest Service officials in charge of the project were in no hurry to try out their new weapon until they were sure of it. Fate helped them decide. Thunderheads marched across Region One on legs of lightning, and wherever they trod they left a fiery footprint. The summer of 1940, exactly thirty years after the Great Idaho Fire, had the highest number of lightning-set fires on record to that time.

The first fire jumps in history were made on July 12, 1940, by Rufus Robinson of Kooskia, Idaho, and Earl Cooley of Hamilton, Montana, on the Martin Creek fire in the Nezperce National Forest. Counting their training leaps, the twelve men of history's first smoke-jumper squad made ninety-nine jumps to nine "selected" fires. The quotation marks mean that those fires were in the most remote and difficult-to-get-at places in the Region. Crews traveling overland would have taken days to get there. By the time they did, the fires would have been out of control. The smoke jumpers pounced upon them from the air and smothered them while they were still small.

Two other significant events took place during that dramatic summer. Three days after Robinson and Cooley made their historic leaps, smoke jumper Chester Derry scored another first, a successful descent on a rescue mission to a crashed airplane in the Bitterroots. He was the fore-

. S. Forest Service

Smoke-jumper Rescue crew going down-hill with Stokes litter. Rescue teams such as this have maintained an evacuation speed upwards of $4\frac{1}{2}$ miles per hour.

Smoke-jumper Rescue crew transporting patient in Stokes litter. All hands are required to overcome some obstacles.

. S. Forest Service

runner of those pararescue teams whose exploits on land and sea, in peace and war, are now famous.

The other event was less dramatic but had results just as far-reaching. In June of 1940, four U.S. Army staff officers visited the smoke-jumper base at Ninemile. They prowled about for days watching, taking pictures, and filling their notebooks with sketches of the training aids. They told very little about their purpose but asked innumerable questions. They were, in fact, exploring a new dimension in warfare, the paratrooper.

The leader of the group, Major William C. Lee, went from Ninemile to Fort Benning, Georgia, where he founded the Army's Parachute School, the giant military offspring of the Forest Service's Aerial Fire Control Project. Major Lee later commanded the 101st Airborne in the World War II invasion of Normandy. And many a smoke jumper proudly and gallantly wore the emblem of the Airborne.

As for the smoke jumpers of 1940, to go from tryout to operational in a single season was a remarkable feat in itself for a project that involved inventing new equipment, training pilots as well as parachutists, and developing an entirely new system of technique and control. Two men not yet mentioned did as much as anyone to make the experiment a success. These were the legendary bush pilots of the West, Dick and Bob Johnson. Their Johnson Flying Service in Missoula was, as a matter of fact, the only organization in the world at that time with the equipment and the know-how to deliver men by air to a fire. They were another prime reason why the smoke-jumper school was located at Ninemile.

The smoke-jumper corps has expanded season by season, from that pioneer handful of twelve to 425 in 1967. In that year, another bad fire season, the parachutists made 4,440 jumps onto 1,247 fires in all parts of the West. One group of seventeen smoke jumpers from a satellite base in Grangeville, Idaho, by themselves made 257 descents to eighty fires.

U. S. Forest Service

Dick Johnson has done much hazardous mountain flying for the Forest Service

It is the inevitable fate of any new technique adopted by a progressive organization that it no sooner has run the gauntlet from experiment through tryout to operational than something better comes along. There was always one serious drawback to smoke jumping—the long walk back. An airborne firefighter could be over his target, ride his parachute down and squelch the fire all in a matter of hours and minutes. But it might take him days to fight his way out of the wilderness. It happened more than once that calls for smoke jumpers had to go unanswered simply because all these highly trained forest soldiers were somewhere in the back country slowly bushwhacking toward the nearest road or airstrip. Even if the smoke jumper got out fairly rapidly, his precious equipment—parachute, jumpsuit and tools—was out there waiting for pickup and return by hammerhead.

Bob Johnson in cockpit of Ford Trimotor plane

U. S. Forest Service

Now, thought the Forest Rangers, if there were only some way to lift a jumper out of the forest as well as drop him onto it . . . !

That mechanical dragonfly, the helicopter, looked like the solution, a sort of two-way parachute. Forest Rangers began experimenting in 1949 and discovered a few drawbacks. The chopper is low on speed, range, altitude and lifting power compared to fixed wing aircraft. Its cost of maintenance and operation is high. It doesn't like the turbulent air currents so often found in the mountains and around fires. But it can still land practically anywhere there's room to swing the rotor. It is the ultimate in mobility short of the antigravity belt. If nothing else good came out of the war in South Vietnam it did breed a whole new family of helicopters with superior performance for the Forest Rangers to take over.

One is struck by the parallels. Out of the shambles of 1910 and 1919 came the network of roads and trails and Evan Kelley's super mules to travel on them. Out of World War I came air transport for cargo and parachutists. World War II produced another breakthrough for the Forest Rangers.

Earlier in this chapter it has been mentioned how smoke jumping got its start in 1939 when water-bombing of fires turned out to be a failure. The aircraft of that day simply did not have the lifting power, but the Forest Service never really gave up the idea. In the vastly improved aircraft developed during World War II they saw at last the means of delivering a crushing jolt of water or chemicals. There were so many things an aerial tanker might do: cool a hot spot on the point of blowing up; hold a break in the line; fill the gaps along the ever-lengthening flanks of a moving fire; knock down an isolated spot fire.

Among the countless types and sizes of warplanes that cluttered military airfields at the close of World War II, the Forest Rangers found a number that had promise as aerial

U. S. Forest Service

F7F air tanker drops 700 gallons of borate slurry on a fire in light timber

tankers. The Navy's great long-range patrol bomber and scout ship, the PBY, could lift two thousand gallons of water. In the end it proved to be too big and clumsy to get in and out of tight corners. The B-25 could carry up to a thousand gallons but moved too fast for hedge-hopping in the mountains seventy-five feet above the tree tops.

One thing the Forest Rangers had learned in earlier experiments was that in order to do any good a water bomb has to be practically shoved down a fire's throat. Otherwise the water turns to mist and evaporates before ever touching ground.

Eventually the foresters found what they wanted in the plane whose name will ever stand for incredible feats of heroism in the Battle of the Pacific: the Navy's Torpedo Bomber. The Grummann TBM is a rugged ship, built to stand up to rough landings on the deck of an aircraft carrier—or, as it turned out, on the bumpy airstrips of the firefighters. Its strong back can lift six hundred gallons—two and a half tons—of water or chemical retardant. Its stout legs can speed it to a fire at two hundred miles per hour and then slow down to a hundred for the drop. It is maneuverable—able to whip into a narrow canyon, make the drop, and then climb out.

Bell Helicopter laying fire hose from the air

S. Forest Service

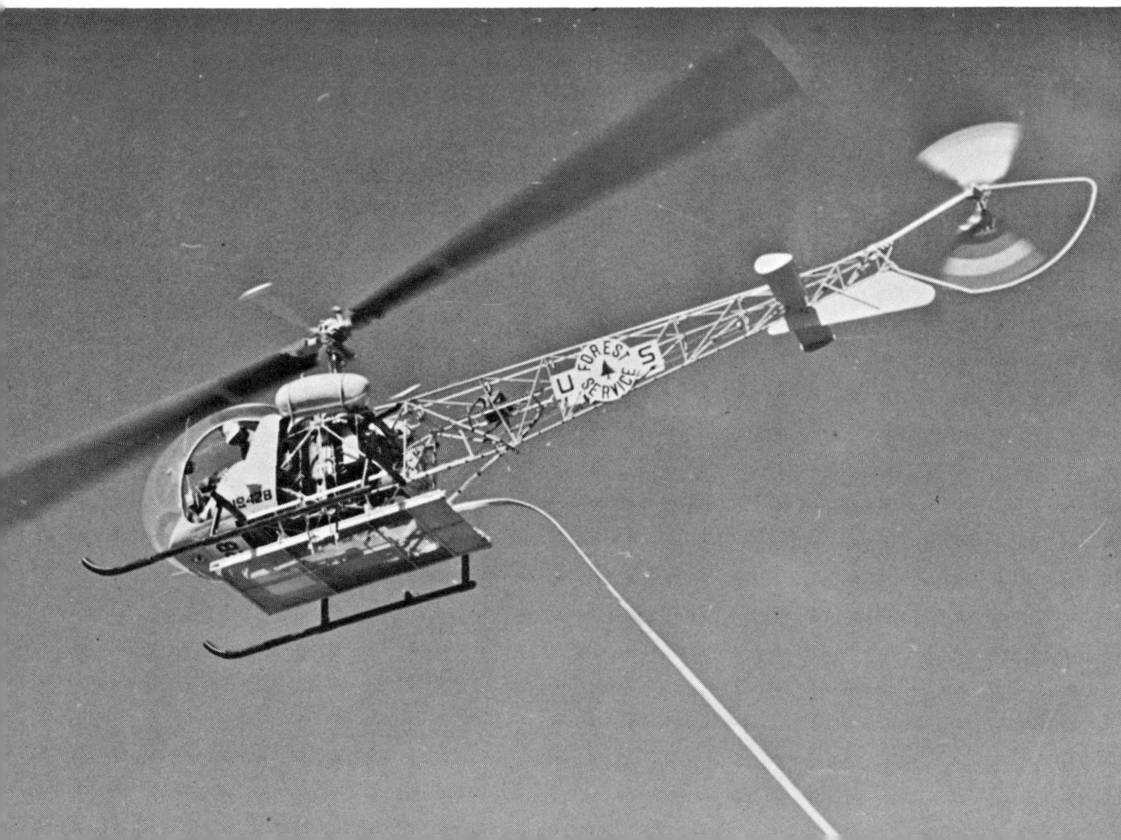

The TBM was in action during the Battle of Camptonville, to be described further on in this story, along with the F7F, an Army fighter-bomber of superior performance that came along just too late for World War II.

Meanwhile the smoke jumpers aren't turning in their parachutes just yet. For that matter, the hammerheads have by no means vanished from the back country trails. Roles and emphases change as new ideas crop up and new techniques are developed. The Forest Rangers continue to adopt any new invention that helps to solve their basic problem of time and distance. If none turns up, they'll invent something themselves.

6

The Battle of Camptonville

Time and distance do not change. A mile still has the same number of feet and an hour the same number of minutes as in Pulaski's day. What changes is man's ability to go farther and do more in the same amount of time. Big Ed should have been at the Battle of Camptonville. It was almost a carbon copy of the situation he faced in 1910, but with a different ending.

Fire years seem to have a sort of rhythm, coming more or less in multiples of ten: 1910 . . . 1929 . . . 1940 . . . 1949 . . . 1959. Half a century after Pulaski's ordeal in Idaho, northern California became the battleground. Although the summer was hot and dry, the Forest Rangers got through it with no serious trouble. But the fall rains, which usually send the red enemy into winter quarters, did not come. October was as dry as the previous months. Day after day the rangers scanned the weather maps for that low pressure area in the North Pacific that would bring relief to the parching forests. Monotonously the weather forecasts repeated: "Continued hot and dry with light winds."

Fire patrolmen traveled their beats around the clock, along the paved highways, up the dusty logging roads, over the back country trails. Scout planes circled overhead like hawks. Lookouts probed the hazy canyons with their ten-mile eyes. The big yellow signs went up everywhere warning tourist, hunter and camper:

EXTREME FIRE DANGER
BE VERY CAREFUL

But small children can't read signs; sometimes neither can their parents. Two toddlers out on a picnic wandered away from the campground. They had a box of matches, and fire has a fascination for children. It was fun to hear the match-heads pop, to see the bright sparkle, to blow it out. Presently a match, still alight, fell onto a clump of bone-dry grass. They couldn't blow out that flame. In a matter of minutes a whole hillside was on fire. The rangers came with bulldozers and tank trucks and pulaskis and put it out.

Ten miles east of Sacramento, California's State Capitol, lies Roseville, on Interstate Highway 80. After it was over, investigators ferreted out what happened there. The tourist zooming along the freeway probably didn't realize that he was passing through one of those foothill towns scarcely to be seen amid the trees and shrubs and flowers. The tourist flipped a cigarette out the window of his car. The glowing coal bounced off the pavement, over the bank and onto a handful of leaves. It was only a coal the size of the end of the tourist's little finger. But this pinpoint was hotter than the melting point of lead. The dry leaves shriveled and charred and smoldered. By the time the first serpent of flame crept out of the leaves and into the tindery bushes, the tourist's cigarette stub had been consumed completely and he was far away. Forest Rangers and volunteer fire department men hunted the blaze he set among the shade trees and flowers and put it out.

Another car drifted along a back country road. To the girl lookout perched in her tower it seemed to be behaving strangely. It stopped, the driver got out and looked around. He climbed back in and the car moved on a few hundred feet. With eyes already red-rimmed from strain, the girl peered through her binoculars. The car stopped again, at a place where the driver could see at least half a mile along the road in either direction. He was getting a can out of the trunk, pouring something over the litter alongside the road. He stooped and his hand made the unmistakable gesture of striking a match.

Firebug!

The lookout spoke into her microphone.

Ten miles away, on a different road, a green pickup truck with the Forest Service emblem on its door made a 180-degree turn. In the opposite direction, a red truck of the California Division of Forestry accelerated so suddenly that its radio mast lay flat over the roof. A sheriff's car howled through traffic on the freeway, shot off on a dirt road.

The foresters put the fire out. The sheriff put the firebug where he would make no more trouble for some time to come.

October 31st. There should have been a sprinkling of snow on the highest peaks of the Sierras. People should have been checking over their ski equipment. Surely the rains would come now. Instead, the weather forecast was even more ominous: "Continued fair with unseasonably high temperatures and low humidity. *Northeast winds, 30 knots.*"

To the anxious men in National Forest headquarters and Ranger Stations, this desert wind was a final blow. From the baked plains of central Canada, across the salt flats of Utah and the alkali beds of Nevada, came the wind, an invisible river of heat that sucked the last drop of moisture out of every growing thing. Even as the first gusts whistled through the passes and down the canyons of the Sierras, the Forest Rangers were sending out the red alert.

Work parties in the field checked in by radio every hour. To logging companies and highway contractors, the red alert meant "Be ready; we may need your men and your equipment at any moment." Truckers, bus companies, food supply houses, and contract airlines were notified. If a firefighting army had to roll, its manpower would come from military posts, farm labor pools, prisons, schools, employment agencies, and the corner of Broadway and Main. In each Ranger District, the dispatcher got out his list of local cooperators—men who had other jobs but who had taken training in firefighting—and put them on stand-by. In Forest Service headquarters a thousand miles from the Sierras, clerks typed the rosters of "overhead," the specialists of a modern firefighting organization: sector bosses, line scouts, equipment foremen, supply experts, timekeepers, aerial scouts—known as bird dogs—and a host of others. Fighting forest fire had become a highly technical and mechanized craft. Through its network of telephone and radio the Forest Service put into effect plans long since made ready. No man moved or stopped his usual occupation, but every man on those rosters of "key personnel" had his kit packed and never strayed far from a telephone or a radio. Those bone-dry forests of northern California were like a magnet whose lines of force were felt in every direction.

On the Plumas, Tahoe and Eldorado National Forests there was nothing to do now but keep up the ceaseless patrols to try to prevent the holocaust—and wait. Men had done everything possible to head off calamity, as well as get ready for it. In the end, it was man, or at least the works of man, that lit the fuse. In a canyon far back in the Tahoe National Forest that violent desert wind blew a tree over a cross-country powerline. From the broken, short-circuited wire, man-made lightning struck the parched earth. Yellow flames sparkled in the weeds and grass, jumped from them into waist-high bitterbrush and manzanita, leaped from them into the branches of pine and fir. Flames turned from yellow to red as they gorged on the oily needles. Overhead bloomed a gigantic umbrella of smoke.

S. Forest Service

Goose Creek fire, Idaho National Forest, Idaho, August, 1940

Cat making fire lane near edge of Robie Creek fire, 1955

U. S. Forest Service

This was historic ground, known first to the trappers and traders, then to the settlers and miners of pioneer days. By 1959 it had come to be the center of a thriving lumber industry, of cattle and sheep grazing, of homes surrounded by orchards and flowers, of the summer and winter playgrounds known the country over.

From opposite sides of the forest, two Tahoe rangers, Lynn Horton and Jack Woolfolk, raced their pickups over the dusty logging roads with the few helpers they could collect on the way. They would have been in time to catch any ordinary fire while it was still on the ground and stamp it out.

Not this fire. With the northeast gale at its back, it was already throwing banners of flame fifty feet into the air. It burst through a pass and then like a tidal wave rolled westward down a heavily timbered canyon. Grimly the rangers spoke into their radios. The fire was out of control.

Fifty miles away, at Forest headquarters in Nevada City, the fire dispatcher had already guessed as much. With a word into his microphone he turned those long-made plans into action.

Within a radius of a hundred miles two other major forest fires were already burning. Disaster on the scale of 1910—or even worse—was in the making. For this area of northern California, which had once been a wilderness much like Pulaski's Idaho, was now dotted with homes and lumber mills, farms, villages and cities. Among the villages was Camptonville, headquarters of Lynn Horton's Ranger District.

Camptonville was no stranger to forest fire. Twice in a lifetime that went back to Gold Rush days it had been burned out by forest fires. (As a minor footnote to Western history, Camptonville was the place where a frontier blacksmith and prospector invented the Pelton waterwheel, which revolutionized the art of converting falling water into useful power.) The year 1959 could well be Camptonville's third burn-out, for the town lay directly in the path of the flames.

S. Forest Service

Robie Creek fire burned for 5 days and covered 7000 acres

The fire was still twenty miles away, but while husbands, fathers and brothers reported to ranger headquarters, the wives, children and sisters of Camptonville began loading their cars.

In the old days, a ranger would have been on his own, fated to do the best he could with what he had. In 1959, the first wave of reinforcements arrived almost before Horton finished his report. Lumber mill and logging crews came by bus and truck and were quickly put on the fireline by Horton's staff. Behind them, drivers expertly tooled their many-wheeled transports along the narrow roads, each trailer sagging under the enormous weight of a bulldozer. From Civil Defense came pumpers with a thousand gallons of water sloshing in their iron bellies. Buffeted by vicious air currents, a helicopter slurruped over the doomed valley while its bird dog plotted the location and rate of spread of the fire.

In Camptonville, smoke drifting ahead of the fire turned noonday into evening. Already the town was without power, for the same lines that served the town were the ones that had touched off the conflagration. Straddled by the fire, that powerline now lay in ruins, wires melted and poles burned off for miles. But Lynn Horton's normally quiet Ranger Station throbbed with energy. A command group of fire specialists had moved in. Their portable power generators racketed in the yard. Telephone men were stringing extra lines. Radio masts sprouted from the roof. In a field nearby, tents sprang up; trucks unloaded mountains of food and equipment; an open-air kitchen set a lavish table, for fire-fighting is hungry work. The main body of firefighters was coming in by the busload, from military posts, contract labor pools and recruiting points in the cities of the plain. Some

U. S. Forest Service

Green River Fire, California Community Threatened! Above the community of Silverado flames 50 feet high roar down to the 80-foot-wide fire break.

U. S. Forest Service

A helicopter heads to a forest fire with a large 450 gallon water container made of fiberglass dangling from its underside

of the overhead—Forest Service language for fire specialists—had been flown in from as far away as Colorado.

It was an astonishing display of organization and power, even in small details. From Regional Headquarters in San Francisco, over two hundred miles away, a courier delivered a bundle of special maps so new that they were still limp from the heat of the duplicating machines. A meteorologist set up his mobile weather station, for in this fight the variation of a few degrees in temperature and in humidity and a few miles per hour in wind velocity would guide the fire boss in his decisions.

The Ranger Station was a madhouse of blatting radios, jangling telephone bells and milling people. In one room there was relative quiet. Here Keith MacDonald, the fire boss, huddled over a map table with his plans and service chiefs. The service chief told him what he had to fight with as of that moment: six bulldozers, thirty-two pumpers, nine airborne tankers, and a thousand men, all deployed in that backwoods area in less than forty-eight hours. The plans chief, Paul Madden, told the fire boss what he had to fight. His grease pencil traced on the map where the fire was by latest report from the scouts, traveling on how wide a front and how fast. Without comment, Madden penciled in the location of the most favorable places to make a stand.

Upon a fire boss lies the crushing weight of decision. If he commits his men and equipment in the wrong place, or at the wrong time, the fire may get away completely, and firefighters may pay for his mistake with their lives.

All the long, hot day MacDonald had been deploying his army along the flanks of the fire, to keep it from widening its front. No human power could stop the forward charge of the red enemy while it was riding the gale. MacDonald tried once, where a logging road cut across the canyon from side to side, a ready-made defense line. Bulldozers widened the strip of bare earth. Snarling chain saws ripped through trees and brush. Ground tankers dampened the woods. Aerial tankers plastered the fuels in front of the fire with chemicals. The idea was to hold the fire on the ground where the pulaski men could strangle it with dirt.

The fire dragon, leaping from treetop to treetop on the wings of the northeaster, was not to be denied. The situation became very tricky. Because of some freak of turbulence in the air torrent flowing down-canyon, an eddy developed behind the defense line—a pocket of superheated gas. It was the condition for a blow-up, that dreaded phenomenon of very hot fires where acres of timber can explode into flame simultaneously, from a single spark. Some fire-wise crew boss detected it and passed the alarm by radio, and MacDonald pulled his men out of the trap. On a front a mile wide and fifty feet high, the fire rolled over the first defense line.

Along the flanks the stubborn holding action went better. Without the wind at its back, the fire climbed the sides of the canyon at a crawl. On the ridge crests, the firefighters held their lines. Spot fires broke out behind them where the red enemy lobbed embers over their heads. But the roving patrols were there to pounce upon these young fire dragons, choke them with dirt, drown them with water, and cut them to pieces with pulaskis and other hand tools.

In late afternoon the sun glared through the smoke like a red-hot cannonball. Then it dropped out of sight. A degree at a time the temperature began to fall. The flanks were holding. Still out of control, the front of the fire was approaching a larger canyon at right angles. It was another and better defense line with a watercourse at the bottom of the new canyon and a main highway halfway up the opposite side. It was also a place where the fire might make an end run around the flanks, cut the army in half and break out on a broad front. In the corner where the two canyons joined, Camptonville awaited its destiny.

The fire boss had another decision to make.

The alternatives were on the map table before him. Beyond Camptonville west, on the path of the fire, lay an almost roadless country, badly jumbled, without another favorable defense line for fifty miles. Plans chief Madden was already struggling with the time and distance problems of redeploying the whole army. By now, MacDonald had been on the job twenty-four solid hours, sifting the flood of information that poured in by radio, telephone and messenger,

S. Forest Service

Dust is one of the best cooler-downers known

A CCC boy fights the Dunsmuir bridge fire, October, 1933

S. Forest Service

making the decisions that only he could make. He knew that his mind was beginning to lose its cutting edge.

On the ridge above town an equipment boss, Nels Stone, spoke into his radio. He was shepherding a fleet of bulldozers along the flank of the fire, trying to turn it away from Camptonville. But the fire was outrunning him.

"This isn't going to work," he said. "I'll have to drop out and pick it up again at the edge of town."

"Put a crew in behind him to fill the gap," ordered the fire boss.

Someone remembered that down in the bottom of the main canyon there was a single house. The owner was on the fireline. His mother was in the house alone. She had no telephone.

Seeing the look in MacDonald's eye, the man who brought that message didn't wait for an order. He went and got the firefighter's mother.

As evening came, the ridges east of Camptonville were haloed in orange. The fire was only a mile away now. Individual trees were silhouetted as flames sparkled on each twig and branch. The people of Camptonville stood in the streets and watched, beside their loaded cars. The windows of their houses were dark, the doors unlatched. Why lock a door that fire would open with its own key?

MacDonald pondered. Suppose he made a stand at the canyon and lost it? He could never recapture the time and work expended, and the fire might outrace him to the next defense line. Suppose he gave up here and used the precious time to redeploy? By then maybe the murderous wind would drop. That second choice was tempting, the choice of caution. The plans chief fidgeted. If they were going to pull out, they'd best be at it, in time to move all the rolling stock before the fire cut the highway.

A scout halted his jeep on the highway and peered upcanyon at the fire. It was hard to be sure with the smoke stinging his eyes, but he thought that the fire had slowed down. There was more smoke and less flame. The glare reflected from the smoke cloud was dull red instead of bright orange. Along wide sections of the front the fire had stopped crowning and was advancing along the ground. The wind no longer plastered his shirt to him like another skin.

The scout said into his radio, "She's coming down out of the treetops. Maybe it's only local but where I am the wind has dropped at least ten miles per hour."

"Check with the forecaster," MacDonald ordered.

Back came the report: "In the locality of the fire, the wind has dropped to twenty knots. The condition may be expected to last through the night. Gale-force winds again tomorrow."

MacDonald made his decision: "We'll stand on the line of the highway."

Operators began to chant into their radios: "Sector Boss Three, this is Base Camp . . . Tanker Boss . . . Tractor Boss . . . Camp Boss . . . we're standing on the highway."

Headlights flared in the darkness. Truckloads of men rolled past the equipment stacks, picking up bundles of tools on the fly. Pumpers trundled down the steep exit road from Camptonville and onto the highway. The monstrous bulk of a bulldozer squatting on its lowboy drifted through the smoke. Specialists in jeeps darted by the larger vehicles with their chain saws and propane torches.

There was a nervous stirring among the townspeople. This looked like a pull-out. But no one actually got into a car and left. These were people who lived close to the forest and close to foresters and not all the signs said Go. The headquarters wasn't packing up. They knew better than to dash out onto the highway and create a traffic jam. When it was time to go Ranger Lynn Horton would tell them.

On his situation map Madden extended the solid black trace of fireline made and held into the dots and dashes of line to be built. With pencil poised he asked, "What about the town?"

This was no moment for sentiment or grieving over inevitable destruction. They must think only of destruction that might be prevented.

"Fight for the town," replied MacDonald. "Throw a dozer line around it. Back it up with those Civil Defense pumpers. Notify the people to keep the streets clear."

Over Camptonville, the curtain of light changed from orange to red. Nels Stone's bulldozers followed their headlamps around the perimeter, uprooting grass, brush, trees and flowerbeds, leaving behind them a twenty-foot swathe of naked, fireproof earth. A whole fleet of pumpers sloshed along the streets, deployed so close together that they could throw up a virtual wall of water. The tide of fire advanced on them crabwise along the hillside.

Down on the highway, men and machines were placed and ready. Below them they could see the enemy, an entire valley of fire. For the moment there was nothing they could do but watch. Here the wind was gusty. There were even downdrafts. Men with propane torches moved along the embankment setting backfires to rob the main fire of its fuel. All sounds were muffled—the coughing of smoke-irritated throats, the idling of engines, the sullen roar of the fire.

The fire, which always travels faster uphill, began its charge, not on a solid front, but cunningly, in individual rushes, under cover of waves of heat and smoke. All it needed was a single foothold beyond the highway from which it could spread and multiply. As each charge reached the highway, the firefighters met it hand to hand. They quenched it with water, choked it with dirt, physically pushed it back down the mountainside. The glare was blood-red, and the men seemed to move in slow motion as if swimming in a river of blood.

A dozen times the fire, balked at the highway, fired showers of embers and flaming brands over the heads of the defenders. These lodged at the top of the embank-

A fire-fighting plane

ment and smoldered briefly while they fed on superheated air and cured grass. Then the flames stretched their hungry tongues toward brush and trees. The Battle of Camptonville could have ended in defeat right there. But many of the pumpers were equipped for fighting city fires. The engine men quickly set their ladders against the bank. They stormed the vertical walls. They threw the fire bodily back onto the highway, and sullenly it died there.

Except in one place. One single ember, riding a gust of wind, flew a quarter of a mile before it plunged to earth. All night, unknown to anyone, it smoldered and fumed. Just as the forecaster had promised, gale force wind came with daylight and fanned the fire into roaring life. But it was too late. The bird dog was aloft. He spotted the new fire through its camouflage of smoke. He called to the fleet of aerial tankers orbiting overhead. One by one they peeled off and loosed their loads of chemical. They literally drowned the spot fire.

So ended the Battle of Camptonville.

7

Aftermath

The town of Camptonville was saved. So were the uncountable thousands of acres of timber and forage, of wild creatures, watershed and recreation area that would have been destroyed if the Forest Rangers had not knocked out the fire.

The firefighters went back to their military posts, to farm and orchard, to factory, mill and school. The teams of "overhead" flew back to their distant stations. The great array of bulldozers, tankers, trucks and aircraft dispersed as rapidly as it had assembled. In the town itself, people unloaded their cars and began talking about planting grass and flowers to cover that fire-moat of raw earth built by Nels Stone's tractors. The fire was out, but it had left a gaping wound twenty-five miles long by two miles wide.

What happened next is not as exciting as the Battle of Camptonville but every bit as important. In the early days of the Forest Service, the rangers had neither the know-how nor the manpower and equipment to do anything about a burn. Nature had to heal the wounds. First came the weeds; then the grass and wildflowers; then the brush; and finally the seedling trees, pushing their crowns up toward the light a few inches a year. If all went well, a forest might rise from its ashes in a century. Meanwhile the ghost forests stood, some of them from as far back as 1910, the skeleton trees toppling over one by one like monuments in an abandoned graveyard.

There are several things Forest Rangers don't like about Nature's method of reforestation. It takes too long; too many things can go wrong with it; and in the natural cycle, the first trees to appear are often of poor quality or type. This intermediate forest has to mature and die out before what the rangers call a climax forest can take its place. A climax forest is the highest type that the particular soil and climate can support. In some cases the growth of brush following a fire becomes so dense that trees can't force their way through it. Robbed of moisture, food and sunlight, they take root, then wither and die. Finally, in a very hot fire, all vegetation is devoured, including the mat of leaves, twigs, needles, grass and other litter that fertilizes the soil and acts like a layer of blotting paper to hold moisture. Within that mantle of vegetation also sleep the seeds from which new plants can spring, and they too are destroyed.

When the rains come to such a burn, they wash away the unprotected soil, carve gullies into the hillsides, and choke the streams with mud. The fish die. With nothing to eat, wilderness creatures great and small must migrate or starve. The once-verdant country becomes a desert in the sense that no forest grows again without help, or at least not for a very long time. I once lived in a northern Montana valley where the 1910 fires burned their hottest. On the mountainsides across the Middle Fork of the Flathead River, in what became Glacier National Park, there were no second-growth trees thirty years after the fire. I revisited the place fifty years after the fire, in 1968, and a few scrubby jackpines were at last taking hold. Among them stand the gray skeletons of the lordly tamaracks that are the climax forest of this area. On my side of the river, where the fire didn't burn so hot, natural reforestation is well on the way. But by natural means, two centuries will scarcely be enough to bring back the tamaracks.

Long before the Camptonville fire, the Forest Rangers began doing something to help Nature. To begin with, they could at least stop the appalling waste that results from letting fire-killed trees simply rot. If the trees can be harvested within a year, before insects and decay get into them, they can be turned into useful products. At Camptonville, the timber cruisers were at work before the burn stopped smoking. A timber cruiser is a man who measures a forest. With cruiser stick and tape and log tables he determines how many board feet of lumber it contains. A "board foot" is lumberman's language meaning one square foot of wood an inch thick.

The cruisers found that there was enough lumber in the Camptonville burn to build ten thousand houses worth over a million dollars. The Forest Rangers called in the loggers and sawmill operators, many of them the same men, with the same equipment, who had just finished fighting the fire. Bulldozers snorted again, building roads instead of fire line. Chain saws snarled through charred bark into wood that was still fresh and clean. In a crash effort that lasted all winter they salvaged every board foot.

Salvage was only the beginning. Along with the timber cruisers, other Forest Service specialists swarmed over the area. The erosion control teams ran their contour lines. Bulldozers followed the survey stakes, building level terraces and ditches to check rainwater and to hold the soil in place so that the first storm could not wash it into the stream bed. Plant men tested the soil and studied the rainfall and temperature records. Orders went out to seed companies and tree farms. By hand and from the air, grass seed was scattered, to sleep through the winter and sprout with the thaws of spring. These quick-growing plants would grip the soil with the threadlike fingers of their roots and so check erosion. At the end of their one-summer life span, their stalks would carpet the land and start the long process of refertilizing the scorched earth. During the spring and summer following the fire, planting crews set out thousands upon thousands of young trees, not haphazardly as in a natural reseeding, but properly spaced so that each would have its proper share of moisture, food and sunlight. What Nature might do by herself in a century she can do in half that time with the help of the Forest Rangers.

Reforestation is only one of the ways in which rangers conserve our timber resources. Man and fire are not the only enemies of the forest. A tree is a very complicated organism. Like every living thing on this planet, it has its insect pests and diseases. They attack it through the leaves and needles, by which it breathes; through the roots, by

Rehabilitation crew plants Ponderosa pine on the Escudilla Mountain burn of June, 1951. Forty-four million feet of damaged timber were salvaged following the fire.

S. Forest Service

which it eats; through the bark, which is like a human being's skin; and through its circulation system. The most destructive pests are immigrants from foreign lands. In one way or another they sneak into this country, leaving behind the natural enemies that helped to keep them in check. We know from history what a new and strange plague does to humans—what smallpox did to the Indians, for instance. Trees are no different. For this reason, the Dutch elm disease has almost exterminated this stately favorite of New England.

Fighting insects and disease is not as spectacular as fighting fire, but it is no less important. Often the same equipment is used. The plane that was dropping chemicals on a fire last week may be spreading insecticide over square miles of forest later on in the season. And in many other ways the Forest Rangers battle the tiny enemies—some of them too small to be seen with the naked eye—that can kill a mighty tree.

Sometimes, when an epidemic breaks out, many acres of forest must be cut down, the bark and leaves stripped off and burned, and the logs sprayed with pesticide. Scientists patiently study the habits of these pests. In some cases they have found that the disease spends part of its life in a certain kind of grass or brush that acts as a host plant. If the host plant is grubbed out, the chain is broken and the disease dies. Other scientists study the natural enemies of the pests and can often bring them in to act as the foresters' allies. Plant detectives scour the world for trees that have built up a natural immunity to their disease and insect enemies.

You might think that a Forest Ranger has finished his job when he has planted a tree, protected it from fire and other enemies until it has grown straight and tall, and has seen it finally hauled away to the mill to be turned into the endless list of useful products that can be made

J. S. Forest Service

3B1 Model helicopter belonging to Johnson Flying Service of Missoula flies timbers into the North Fork of the Blackfoot River to complete a trail bridge taken out by a flood

A B-17 spray plane laying down a 600′-swath in spruce budworm infested timber in the Targhee National Forest

J. S. Forest Service

from wood. But one of the most important ways of conserving our forests is to get the best use out of them.

We live in an age of metal and plastics, but man has yet to find or invent anything as useful as a tree. From hammer handles to railroad ties, from toothpicks to shingles, from furniture to floor joists, from paper to bridge planks, from turpentine to telephone poles, a tree is a warehouse of products. Wood can be sliced, sawed, planed, fitted, laminated, shaped or carved into any form a man desires. It can be fastened with nails, screws, bolts, staples or glue. It is durable, strong, light and beautiful. It can be painted, stained, polished, waxed, varnished or left natural. From the roof over your head to the fire on your hearth, wood is a friend to man. If a man sat down to make a list of all the good things that come from a tree, including its shade on a hot day and the fruit from its branches, he would be busy for a long, long time.

Early in its history, the Forest Service realized that the manufacture of wood products was a very wasteful process. Consider a tree that is to be turned into lumber. The limbs, top, stump and roots are discarded before the tree leaves the forest. At the sawmill, the first thing done is to make the round log square by slabbing it off on four sides. These slabs, and the bark, are thrown away or burned up just to get rid of them. Next, the squared log is sliced into rough boards, and the sawdust from each cut is waste. Finally, before the boards are ready to go to a lumberyard, they have to be planed to make them smooth and uniform in width and thickness. The shavings are waste. Thus the long career of a tree from the forest to lumber in a house is strewn with waste every step of the way, and the finished product may be only half the original tree.

Next to building material, the most important wood product is paper. In order to make paper, a tree is stripped of its bark and then cut up into chips. These chips are cooked with chemicals at high temperatures in order to break down the lignin, which is Nature's glue, and release the wood fibers. There is still more processing before the wood fibers become paper. Although papermaking is not as wasteful as lumber manufacturing, the loss may still amount to as much as one-quarter of the tree as it stood in the forest.

Reducing waste in the manufacture of wood products is one obvious way to conserve our forest resources. In 1910 the Forest Service went to work on the problem by setting up the Forest Products Laboratory in cooperation with the University of Wisconsin. The task of this now world-famous laboratory has been to study wood and how to make the

The Nebraska tree planter showing a row-marker attachment. The attachment is made of telescoping pipe which can be adjusted to the desired row width.

. S. Forest Service

U. S. Forest Service

District Ranger Ed Corpe shows two Boy Scouts how trees are planted in Southern California's National Forests

The proper techniques of tree planting are explained to farmer Hugh Jones and others. Ten thousand slash pine seedlings were planted on 13 acres of corn land in one of the first 1957 plantings under the soil bank program.

U. S. Forest Service

most of it. One of its important contributions was the development of better saws and better ways of using them. Attacking the problem from the other side, the Forest Products Laboratory has also found ways of turning sawdust, shavings, and even bark into varieties of useful products ranging from insulating material to plastics and chemicals and decorative paneling.

Lignin, the principal by-product of papermaking, was a special challenge to the scientists. Not only was it wasted, but the paper factories dumped it—along with the chemicals from the cooking process—into rivers, where it poisoned all aquatic life and made the water unfit for use by human beings. Out of lignin the Forest Products Laboratory has extracted a host of useful products, including an adhesive for linoleum and flavoring known as vanillin for ice cream sodas.

Plywood, which is one of man's triumphs in the use of wood, was invented long ago by the Egyptians but was not mass-produced until the twentieth century. To make plywood, a log is peeled by knives instead of being sawed. The thin sheets are then glued together, with the grain running in opposite directions. The result is a panel stronger than the original wood—and no sawdust. For a long time after plywood was invented, it could be used only indoors, where it would be protected from weather. The glue used by man to fasten the sheets together wasn't as good as Nature's glue. If the panels got too damp or too dry, they simply came apart.

Waterproof glues for plywood were developed in Europe in the 1930s. The Forest Products Laboratory has improved upon them until nowadays if a plywood panel fails it is likely to be the wood itself that has pulled apart. Nowadays practically an entire house—everything but the framework—can be built of plywood at great savings in time, labor and material. Plywood also revolutionized the art of building boats. There is probably nothing made by man that takes a worse beating than the hull of a racing boat, from tiny outboards to those roaring hydroplanes that skim the water at over a hundred miles per hour. And they are built of plywood.

Although it is a familiar enough sight on any construction project, another use of this amazingly versatile building material may come as a surprise. As everyone knows, concrete is a mixture of water, sand, gravel and cement. The builder can pour it in any shape he fancies. But until it hardens, concrete has to be held in place by a form. For this purpose, waterproof plywood is much better than ordinary lumber. It is stronger, lighter, smoother, has fewer joints, and is faster and easier to erect or take down. It can be used over and over again. Plywood has helped to make it practical to build most of our big structures—bridge abutments, skyscrapers and dams—out of concrete instead of brick or stone masonry.

At every stage in the management of a forest, the technician and specialist work side by side with the ranger. The results are fewer losses from fire and disease, less waste, more and better products, and forests that will go on producing as long as there is a Forest Service.

8

Children of the Forest

When a ranger protects a forest, or restores one that has been damaged, the results go far beyond the planting of a tree. Our National Forests are the final refuge of the wild creatures that have lived on this continent since long before men came, white or red. Great and small, the forest provides them with food and shelter, from a bear gorging himself on huckleberries to a beaver fashioning his lodge of mud and sticks. A deer browses an alpine meadow; the playful otter cavorts on his mud slide; a mouse lines his nest with silky grass. A forest would be a haunted place without the splash of a jumping trout, fleets of waterfowl navigating the lakes, the lonesome bark of a coyote or the echoing challenge of an elk, the glimpse of a squirrel showing off his acrobatic skill, a grouse strutting for his mate.

It was inevitable that the home territory of these wild creatures would shrink as our nation transformed a wilderness into an agricultural, industrial and urban empire. Millions of acres of forest are gone forever, the land occupied by farms and highways, factories, cities and airports. During this dramatic and often violent changeover, flocks of birds, herds of animals, species of fish, and even races of men were more than reduced in number. Some of them vanished entirely.

To take only two examples, passenger pigeons, which once darkened the sky during their annual migrations, are extinct; and the buffalo, which once roamed the "sea of grass" in millions, now exist only in small numbers in outdoor museums. Passenger pigeons were ignorantly exterminated by market hunters in the old and false belief that the supply was inexhaustible. Destruction of the buffalo, on the other hand, was deliberate. It was done partly to deprive the warlike Plains Indians of their food supply so that white men could more easily take their land, and partly because buffalo robes happened to be an article of fashion. The industrious beaver almost went the same path. It was reprieved only because the time passed when a well-dressed gentleman had to have a hat made of beaver fur.

These are inglorious pages in our history. But the buffalo was doomed in any event, if the United States was to become a nation stretching from "sea to shining sea." The land roamed by this stupid, ugly and quarrelsome beast had a higher destiny in ranches and farms. Perhaps passenger pigeons and other wilderness creatures who suffered extermination were also doomed. Some naturalists will argue that these victims of change lacked the ability to adapt, to find a way of existing under a new set of conditions. The Forest Rangers take no part in the argument. As practical men they have no time to lament over ancient history. They are more interested in helping creatures that were able to adapt and did survive the early slaughters.

Among many examples, one of the best is the elk. In frontier days, the elk was hunted remorselessly, not for his meat, although it is better eating than buffalo; not for his hide, although it makes better leather; but because in his jaw are two teeth that could be turned into elegant watchcharms. It is one of those strange twists of fate that invention of the wristwatch ended the demand for elk-tooth watchcharms and so helped to preserve one of the finest big game animals of North America.

U. S. Forest Service

Red-tailed hawk, sitting on a rock in Kettle Moraine State Forest, Wisconsin

Skunk sniffing the air in Monogahela National Forest, West Virginia

S. Forest Service

The principal reason why elk did not go the way of buffalo is that they had the heart and the head to adapt to a new set of living conditions. Like buffalo, elk are eaters of grass—grazing animals. Deer, on the other hand, are browsing animals; that is, eaters of leaves and brush. This may seem to be a very small difference, but it is an important one. Again like buffalo, elk lived on the plains and in the foothills. In summer, they went up into the mountains to feed on the lush alpine grasses and to escape heat and flies. This was something buffalo never had sense enough to do. But in winter, when grass was buried under several feet of snow, elk moved back to the plains. Deer could winter in the mountains because their food supply of twigs, berries, leaves and bark grew above the snow line.

Relentlessly pursued by man, buffalo could not or would not change their ways. They stayed on the plains and died. But elk learned how to live in the mountains the year round, how to paw through snow to get grass, and how to browse, like deer. Even their mating seasons had to change. And they lived. They are one of the most remarkable examples in wildlife history of survival of the fittest.

Men with rifles are the only enemy capable of exterminating elk. Big, strong, swift and well-armed, they are prepared to cope with any other foe. But in the end men saved this noble wilderness creature. Forest Rangers joined with state game wardens to set aside refuges within the National Forest where the survivors of the watchcharm massacres could exist and grow numerous again. This is a bright page in the history of conservation, but it must never be forgotten that elk made it possible by adapting.

Today there are a number of thriving elk herds. In the northern Rockies, the Sun River herd ranges from Glacier National Park into the Sun Range, that storied ground of the once-mighty Blackfeet Indian Nation. The South Fork herd, of which more presently, lives just over the Continental Divide on the Middle and South Forks of the Flathead

S. Forest Service

Bobcat mascot of the boys at Gold Creek CCC Camp, Okanogan National Forest, Washington

River, in the Bob Marshall Wilderness. Farther south are the Yellowstone herds, one of which roams from Yellowstone National Park in summer to the Jackson Hole country in winter. Others winter on the Gallatin River in Montana and in Sunlight Basin, east of the Park. Farther west, in Idaho, is the Clearwater herd.

Elk have been transplanted to other parts of the West, notably Oregon and Colorado, to establish new herds. The largest of these summers in the Umatilla National Forest in Oregon. To the woodsman there is no finer sight than a band of elk strung out along a mountainside. A wise old cow is always in the lead while another watches over the

A contemplative moose sighted by photographer

U. S. Forest Service

U. S. Forest Service

A Ringtailed Cat (or Cacomixtle) at Livingston, Montana

U. S. Forest Service

Golden Mantled Ground Squirrel (sometimes mistaken for the smaller chipmunk), feeding on grain from sacks left at packers station in Tom Beall Park, Idaho

A raccoon in a tree in Monongahela National Forest, West Virginia

U. S. Forest Service

youngsters in the rear. And no sound is more truly the voice of the wilderness than a bull elk's ringing bugle challenge.

To see how men have helped to save animals from extinction, go with me and Ray West, a Forest Service wildlife specialist, on a winter game patrol through the range of the South Fork herd. As Jack Root stands for the all-round forester—the ranger, man of many skills—Ray West represents another kind of forester, the specialist. I met him while doing a specialist's job myself. I was making a survey of the beaver population in the Marshall Wilderness for the Montana State Fish and Game Commission. My interest in wildlife was to write about it; Ray's was to preserve it and see it prosper. Although they were as unlike in manner and appearance as two men could be, there was a strong resemblance between Ray and Jack Root. They were both supremely competent, each in his own job, and they were both content with their way of life.

The South Fork winter game patrol has been made every winter since foresters noticed that a few bands of elk had chosen the Marshall Wilderness as the place for their last stand. This wilderness, one of the largest in the system, straddles the headwaters of the South Fork of the Flathead River in northern Montana and is part of the Flathead National Forest.

At first glance it might seem that the elk had chosen well. Their million-acre stronghold was surrounded on all sides by a buffer zone, fifty miles wide, of some of the roughest country in North America. The only way men with rifles could get through it was on foot or on horseback. To make the defenses even stronger, the game wardens of Montana and the Forest Rangers had spotted game preserves throughout the area. The wily elk quickly learned about those places where men could look longingly but dared not shoot. The odds were still all against the elk.

In summer, the Wilderness was a paradise for the castaways. The bulls, keeping to themselves as is their custom, sought the windy heights. There they idled through the long days, gorging themselves on an endless variety of grasses and shrubs. Lower down, the awkward, long-legged calves frolicked in the alpine meadows while their mothers kept watch for any predator bold enough to challenge their slashing hoofs. The calves grew fat and strong for the difficult days ahead. Inevitably winter brought a sudden and bitter end to the months of plenty.

The rangers wondered anxiously how these creatures of the plains could possibly survive when all their natural food

Bull elk at Jackson's Hole, Wyoming (Teton National Forest)

S. Forest Service

was buried under a six-foot blanket of snow. Probably, they thought, the full-grown animals had the strength and the endurance. But what about the youngsters meeting cold and snow and hunger for the first time? Unless they too could survive, the herd was doomed.

The principal purpose of the game patrol was to count the herd and so determine whether it was growing, merely holding its own, or going downhill. It was better to do this job in winter, when the elk had congregated in those places where there was food they could get at—the willow jungles, the cottonwood and aspen groves, and the meadows along the watercourses. When I went on the patrol with Ray West shortly before World War II, he already knew the answer. He took with him a game warden and a sportsman, myself, so we could see for ourselves.

Our trek began at Coram Ranger Station at the mouth of the South Fork and ended at Monture Ranger Station, over the divide from the headwaters of the great wilderness river, a hundred and fifty miles from north to south, with side trips up many of the tributaries. It was ten degrees below zero the day we pulled out of Coram, and the snow was dry as ashes under our skis. It took us three days of hard slogging just to reach the edge of the elk range at Spotted Bear Ranger Station, the Wilderness boundary. I thought nostalgically of the summer before, when I had made this same trip in reverse in a canoe.

As the three of us plodded through the frozen wilderness we saw everywhere the signs of an epic struggle. Where ski-shod men coasted lightly over the surface, the elk had to plow chest-deep through the snow. Where men found shelter and food waiting for them in trail cabins, the elk trampled out their frosty bedgrounds and dug up whole acres of snow in search of grass. They had gnawed willow clumps down to stubs an inch in diameter. Pine branches, wood, bark, needles and all, they had stripped off as high as they could reach by standing on their hind legs.

In one place, we noted with amusement, the elk had borrowed a meal from a neighbor. On the riverbank, a beaver had patiently cut down a large cottonwood. It must have taken him days, chipping away with his orange-colored teeth, for the tree was easily three feet in diameter. For him and his colony the bark of this tree would have provided food for many a day. But it was doubtful that they got even a mouthful. The tree no sooner fell than the elk pounced upon it and ate every shred. It seemed very fitting that the hard-working beaver, another animal once near extinction and now making a comeback in the forest, should lend a helping hand to a very distant cousin. Knowing the ways of beaver, I felt sure that this one had an ample food supply at the bottom of his pond. He did the tree-cutting for exercise, and because he must. The beaver's chisel-like teeth never stop growing and must be kept worn down. Incidentally, no one has yet solved the riddle of how a beaver takes a log that naturally floats and makes it sink.

As for the elk, there was very little that men could do or needed to do for them beyond giving them a fighting chance. We could now see for ourselves that they were surprisingly sleek on their unnatural and near-starvation diet. In fact, as West pointed out, they were doing almost too well, multiplying so fast that they were overgrazing their winter range. He was pleased to discover that by a simple trick he had persuaded the herd to spread out over a wider area.

The eaters of grass tend to congregate near natural salt licks. Like all creatures, they need the sour crystals for health. Since there are few natural licks in the South Fork, Ray West had packed in salt during the previous summer and created his own in places where there was an untouched supply of food. These areas were now well populated for the first time since the winter game patrols had begun.

It was a small victory in the long battle to preserve the wildlife of our forests. From this battle the rangers have learned some important lessons. One of them is not to meddle with Nature, or at least to meddle as little as possible. Another is to help those creatures that show the ability to adapt and help themselves, like the elk.

Another adventure, in summer this time, shows man at his very best in his effort to help Nature foster her wilderness children. On a day in August, two foresters stood on the shore of George Lake in the Swan Range of the Rocky Mountains. It was one of the most inaccessible spots in the million inaccessible acres of the Bob Marshall Wilderness. That's why we were there, Ray West and I.

Like so many of the alpine lakes of the Rockies, George Lake was built by a glacier. The enormous river of ice scooped out the catch basin—a mile long, a quarter of a mile wide, and no man knows how deep—right under the cliffs at the top of the range. But even a glacier couldn't shave down the reef of living rock at the exit. When the ice melted and filled the catch basin with water more clear than any glass, the reef remained. The overflow waters of George Lake descend to the canyon floor in hundred-foot leaps, a barrier impassable even to a spawning trout. So the emerald shallows and the blue-black depths were barren.

Our purpose had been to put fifty thousand cutthroat trout into that lake, in defiance of all the laws of time and distance for fish planting. Young trout can be transported in tank trucks—or in cans, by pack animal—if they are not taken too far from their hatchery. The limiting factor is ice, to keep the water at the same temperature as the chilly mountain streams and lakes that are the natural home of the cutthroat. Trout have been dumped successfully out of airplanes, and helicopters are even better for the purpose.

Neither of these methods of back-country transportation would do for George Lake. At that time the helicopter had not been perfected. There was not enough room for a fixed-wing aircraft to maneuver—to dive close enough to

Fifty thousand trout eggs on the trail for George Lake

the water for an air-drop and get back out again. George Lake was a hundred miles from a hatchery, fifty miles from a road, twenty miles from an airfield, and six miles from a trail. I was responsible for choosing it as the site for the experiment. West had given me the job of finding the most difficult fish-plant in the entire Wilderness, and George Lake was the result of weeks of exploration.

On his part, West had been developing a brand-new method that might be the answer to stocking such remote and difficult waters. It was a fish culturist's dream of long standing. Many other lakes and streams were barren for the same reason as George Lake: natural obstacles that neither man nor Nature had been able to overcome. If they could be stocked, they not only would attract the far-ranging sportsman but would also act as fish reservoirs to supply the waters downstream.

West was experimenting with plants of eyed fish eggs. An eyed egg is one about to hatch. If you hold the amber globe to the light, you can see the tiny trout curled up inside, waiting for the moment to break the tough protective shell and begin the life cycle which, from an angler's viewpoint, properly ends on a Royal Coachman at the end of a tapered leader. By packing the eggs in ice, West had discovered, it was possible to keep the eggs dormant (on the point of hatching) for several days. The technique had all the earmarks of a breakthrough, for one hammerhead—even one man—could carry more eyed eggs than an entire truck or pack train loaded with fish in tanks and cans.

So far, results had been very promising. Now the wildlife specialist wanted to try an extreme test of his method. There were two requirements. He wanted the most difficult job of fish-planting that Nature could devise, the reason for all my map-studying and exploration. And he wanted the plant to be in barren water so that there would be no argument over the outcome, win or lose. Those foresters who knew where George Lake was—and many who had worked in the Marshall Wilderness for years did not—shrugged their shoulders when they were told of our plans. At that time and in that place, Ray West and I were considered to be a little crazy anyway.

On the first day of the George Lake expedition, West picked up the eggs at a hatchery and trucked them to an airfield near Kalispell, Montana. On the second day, a Forest Service bush pilot flew West and his cargo to the Wilderness airfield at Big Prairie. I joined the expedition there, having just come in from scouting the approaches to the lake. What I reported was not good. We had hoped to mule-pack the eggs and other paraphernalia to the foot of George Lake falls, leaving us only a short hike to the planting site. But the trail up George Creek was choked by windfalls and avalanche debris. West merely grinned in the amiable way he greeted any news, good or bad, and went on with the preparations. At each stage, the eggs had to be re-iced and picked over to remove any that had spoiled. There were ten trays in the egg case, each tray containing five thousand eggs.

On the third day, the refrigerated case with its myriads of tiny sleeping fish joggled the twenty miles to Shaw Creek smokechaser cabin by hammerhead. At this point we were already two days and several hundred miles beyond the range of conventional fish planting methods. We were still six miles from George Lake, five horizontal and one vertical.

On the fourth and final day, we packed the egg case in ice for the last time. It was "final," and "last" because this was the end of the ice. We wrapped the case in blankets for insulation and lashed it to a smokechaser's packboard. Another packboard held the planting trays, wire, thermometers, tweezers and other implements of the fish culturist's trade. Before dawn we helped each other into the slings of the packboards and took off into the spruce jungle that mantles the floor of George Creek canyon. For a while we were able to follow the blazes of the trail, over or under the windfallen trees, through the tangle of buckbrush and

Ray West preparing a planting tray at George Lake. Each tray contains 5,000 eyed eggs

huckleberry, which must be cut out every year. It was our misfortune to be ahead of the maintenance crews.

Presently the trail vanished entirely under the wreckage of the previous winter's avalanches. After that we followed the line of least resistance. On our caulked boots we tight-roped along fallen tree trunks, if they happened to point in the right direction, sometimes twenty feet above ground. Most of the time we squirmed through alder thickets that caught at our loads with a thousand limber fingers. Occasionally we had the easier going of an alpine park, where we waded hip-deep in bear grass. In the end, we floundered along in the numbing waters of George Creek, our caulks grating on the polished stones. The stream bed made a faster pathway than the jungle, and speed was essential. Time was running out on us. In this case, time meant how much longer the ice would last.

I still have a vivid memory of that hike. The day grew hot and humid, with the feel of a thunderstorm somewhere beyond the ranges. The humidity brought out every deer fly in the canyon, thirsty for our blood. I remember the maddening trickle of icewater inching down my spine. I remember the bruises from tripping over hidden rocks and roots. Always, in a fall, one had to cushion the eggs or the fragile equipment with his body.

Half a mile an hour is top speed for a smokechaser going to a fire in such country. We weren't quite in the class of those iron-legged young trailburners. Nevertheless, we made four miles in seven hours, half a mile above the pace. It was noon when we shrugged out of our packs at the foot of the reef that blocked the upper end of the canyon. To look at it from below, one would never dream that there was room to fit a lake between that giant staircase of rock ledges and the summit of the Swan Range. But the cascade of overflow water leaping at us from ledge to ledge reminded us to have faith in the map.

There was a short conference while we wrung out our

water-logged socks and ate a bar of chocolate. Ray estimated that there was enough ice left to hold the eggs dormant for another two hours, no more. In those two hours we had to scale a thousand feet of lichen-bearded cliffs, reach the lake and make the plant. I'm sure the thought occurred to West as it did to me that it might be better to make the plant right now, in the stream, and salvage that much for our effort. We never discussed the possibility. By now it was all or nothing.

We inched our way up along narrow ledges. When each one pinched out in turn, we shinnied up a fault to the next ledge. At one point, Ray held my caulks against a tiny break in a sheer face of rock while I groped overhead for a handhold. A chunk of granite the size of a washtub flaked off at my hips. I clamped it between my knees. My steel-shod boots were digging into Ray's hands and shoulders. Mildly he asked why I was stalling around. I explained about the rock. By then I had a death-grip on a clump of scrub cedar. West sidled away, leaving me dangling. I unclamped my knees and the rock went bounding down the cliff like a rubber ball.

"If my wife could see me now, she wouldn't like it," remarked West.

Two-thirds of the way up we were completely blocked. To our right the cascade spurted in all directions as it fell a hundred feet sheer onto the ledge where we stood. To the left and overhead there was the vertical wall of granite without flaw or fault. There was only one way out. We marched right through the waterfall and up the other side. We found ourselves slogging along a narrow gorge carved by the tireless waters. We had made it up the main cliff, but what other obstacles might there be in our path, and how far was it to the lake? And all of a sudden, there it was.

For a few moments we simply stood there, unable to believe our eyes. I realized that I was very tired. West

The author in a wilderness dilemma

began fussing with the egg pack. The temperature inside it was rising, he reported. The ice was almost gone and the ordeal wasn't over. The eggs had to be planted in running water. The outlet stream wouldn't do. Most of the hatching fish would be washed over the falls. On shaking legs we dog-trotted along the lakeshore, hunting an inflowing stream. We crossed half a dozen watercourses, all dry now in late summer. But after half a mile of this marathon we came to an ice-cold creek, milky from the snowfields above.

West ripped open the egg pack. I got the planting trays ready, anchoring them so that they would float in midstream. And even as we dumped them in, the eggs began to hatch. The tiny trout wriggled through the mesh of the planting trays and into the fishless depths of George Lake.

This isn't quite the end of the adventure. The elk who bugled in the timber, seeming to say, "We've got to move farther back, fellers; the tourists are here!" was good for a chuckle. How we dragged our weary carcasses back to Shaw Creek cabin is unimportant. But on the shore of George Lake Ray West and Monty Atwater swore a solemn oath that in five years, wherever we were and whatever we might be doing, we would come back and catch a mess of trout.

Fate had other plans for us—World War II, for instance. When that fifth year came round it seemed most unlikely to either of us that we would ever see the mountains and the forests again. In the confusion of war and the years after, we lost track of each other. But ten years later I bumped over the dusty road to Holland Lake, the jumping-off place for the Marshall Wilderness. I plodded over the barrier range to Shaw Creek. I bushwacked up George Creek, scrambled up the ledges beside the cascade, and went through the waterfall. It all looked perfectly familiar—the spectacular cliffs, the slowly receding snowfields, the silvery log jam at the outlet, the emerald shallows and the blue-black depths. There was even an elk on hand to blast his surprise and outrage at me.

I put together my fly rod, mounted the reel, and threaded the line through the guides. I chose a fly, a Royal Coachman. I watched that fly sail out and land delicately on the water. And I caught the first trout ever to come out of George Lake since time began.

That exciting moment was more than a reward for the drudgery of a day ten years past. It proved something. The trout I caught was an eight-incher, obviously not from one of the eggs Ray and I had lugged up there on our backs. He was a descendant, probably several generations removed. He proved that our fish not only had survived, but had also reproduced and multiplied. *That* was our reward.

9

The White Enemy

Protecting wildlife and promoting recreation were added to a Forest Ranger's job list only after the Forest Service had grown up, so to speak. People hunted and fished, camped out and explored as they always had. But to the Forest Rangers, the "wood, water and forage" of Secretary of Agriculture Wilson's famous policy declaration had to come first. Until they were safe there was no point in worrying about anything else.

Increasing use of our forests for recreation was a sign that our nation, too, was growing up, that in spite of droughts, wars and depressions we had the security and the leisure to enjoy our wilderness heritage. Outdoor recreation and the part played in it by the Forest Rangers could be the subject of a book in itself, but a new kind of recreation, unthought of in Pulaski's day, will serve as an example. The story is dramatic because as a result of it the rangers encountered a new and unforeseen enemy, the avalanche.

There was nothing new about avalanches themselves. Wherever the mountains are high and steep and snowfall is heavy, these tidal waves of snow come plunging down. They are dangerous only if people get in the way. By a strange coincidence, the greatest avalanche disaster in the history of the United States took place in the same year and in the same area where Big Ed Pulaski fought his historic battle against forest fire.

Actually, it was no coincidence. The fires of 1910 destroyed many of Nature's avalanche barriers, the dense forests of the Pacific Coast ranges, which had always before held the snow in place. Snowfall during the winter of 1910 was extra-heavy on the Cascade Mountains of Washington. As a result, avalanches took paths where none had ever run before. In the center of one of these new paths lay a siding on the Great Northern Railroad, in Stevens Pass, at a place called Wellington. Just a water tank stood there, and a bunkhouse for railroad workers, beside tracks where trains going in opposite directions could pass each other.

In February a great storm swept in from the Pacific Ocean and engulfed the Cascades. On the last day of the month, three trains that had been stormbound for almost a week stood on the siding at Wellington. There the tidal wave of snow caught them, and ninety-six men, women and children lost their lives.

To the Forest Rangers of that time fire was a recognizable enemy. There was no reason in 1910, or for a good many years afterward, for them to do anything about avalanches. The railroad solved its problem by going underground into an avalanche-proof tunnel. The highways through the western passes were allowed to snow under in those days, and no one traveled over them in winter. The forests of the West slept under their mantle of snow. Winter was a time when a ranger could relax, catch up on his paperwork and sharpen his pulaskis for the next fire season.

The change came gradually. In the East, a few people took up the sport of skiing, imported from Europe. Then some ingenious Yankee invented the rope tow, which took the long walk back out of skiing. The few became hundreds. With a long stride westward in the late 1930s, Averell

Harriman's Union Pacific Railroad developed Sun Valley, Idaho, a complete winter resort featuring the world's first chairlifts. The hundreds of skiing fans became thousands. After World War II the dam burst.

The amazing increase in the number of skiers that began in 1946 has continued each year since and shows no sign of leveling off. One explanation for this phenomenon is that skiing is a sport that anyone able to walk can enjoy without being an expert. Another is that human beings

An avalanche descends

Results of a snowslide in the steep-sided canyon of the St. Joe River, Idaho

. S. Forest Service

seem to have an inborn desire to slide, on anything, and a pair of skis on snow is the best sliding combination there is. Perhaps the most important reason is that skiing gives people an entirely different attitude toward winter. It used to be a season to dread for its cold and its long nights. Snow was something to be shoveled off sidewalks and driveways. The only good thing about winter was that spring was sure to follow. Skiing turned winter into fun.

In the early days in this country, skiers clung to the foothills. But sport fans able to afford the trip were bringing back tales from Europe of long, steep runs and powder snow. The United States had a European-style sample in Sun Valley, and that was just about as remote as Europe to the average person, but it had its effect. As the number of skiers increased and their skill improved, they looked ever more longingly at the highlands of the West, where the slopes were bigger and steeper, the snow was drier and the season was longer. The first truly alpine ski area to be developed in the United States was at Alta, in the rugged Wasatch Mountains of Utah. It had the deep, dry snow and the big open slopes. Within half an hour's drive was Salt Lake City, already a skier town.

During Gold Rush times, Alta was one of the richest mining camps in the West. Avalanches were a problem, as they were and still are at many mining camps. Nature has a way of booby-trapping her mineral treasures with hazards of one sort or another. At Alta, the avalanches snipped off a few lives and destroyed a few buildings each winter, but the miners accepted them as one more danger in a dangerous profession. Then, in the climactic winter of 1874, snowslides demolished the camp and took over sixty lives. Ironically, snow combined with fire to produce this disaster. The snow trapped the men in their cabins and fires from overturned coal and wood stoves did the rest.

By this time the veins of gold and silver ore were almost exhausted. The few survivors fled the place and Alta was abandoned, to be rediscovered half a century later by another adventurous crew, the skiers. Alta's gold and silver are gone forever, but its greater riches of sun and powder snow renew themselves each winter.

Alta was within a National Forest, so the rangers took a long look. This snowy canyon promised to be one of the world's great ski developments. But it was not possible to let avalanches treat people in a public recreation area the way they had treated the miners. The Forest Service stationed an avalanche guard at Alta to watch over the safety of the skiers. It so happened that in 1945 the newly-appointed "Snow Ranger" (a title coined on the spot) was an ex-ski trooper of the Tenth Mountain Division named M. M. Atwater.

I had about the same qualifications to become one of the first Snow Rangers that Big Ed Pulaski had to become one of the first Forest Rangers. I had been to no school of avalanches. Such a thing did not exist at that time in the United States any more than a school of forestry existed in Pulaski's day. I was a professional mountainman and skier. My instructions were simple: Protect the public from avalanches. And, as in Pulaski's case, the enemy gave me a rude welcome.

One day in late December, during my second month of duty, Alta was storm-bound. As the early twilight of midwinter darkened the canyon, most of the skiers were safely in the lodges or in Salt Lake City. All were accounted for but three. In an abandoned mine cabin a thousand feet above the canyon floor, three high school boys had set up camp for their Christmas holiday. They had been warned not to go back to the cabin this stormy night because of the avalanche danger. Unknown to anyone, however, they had gone anyway.

For an understanding of what happened next it is necessary to describe how Alta is arranged. Little Cottonwood Canyon, which runs east from the Salt Lake Valley up into the Wasatch Mountains, is so straight that one can look along it as through a gun barrel and sight on the southern edge of Salt Lake City. At the site of both the onetime silver camp and the new ski area, a canyon comes into Little Cottonwood from the south, Collins Gulch. The ski lifts run up this gulch, while the lodges and highway are on the north side of the canyon, separated from the lift area by Little Cottonwood Creek. The cabin in which the boys had set up for their skiing vacation was on the steep north side of the main canyon, above the lodges.

I was halfway up Collins Lift when one of the boys stumbled into Snopine Lodge, plastered with ice and snow, almost too exhausted to tell his story. On the final leg of the climb to the cabin, an avalanche had run them down. Two had been able to dig themselves out. The third boy was buried. By the time I received the message and got back across the canyon, a search expedition had already started.

It was without doubt one of the most disorganized rescue operation in history. Every living being at Alta started up the mountain. Some of them hardly knew where they were going or why; they were simply following the crowd. Others were not properly dressed or in safe physical condition to attempt a thousand-foot climb in waist-deep snow. As I plodded up the mountain with a rucksack full of flashlights and first-aid equipment, the trail looked like the pictures one sees in history books of Napoleon's Retreat From Moscow. There was no direction for them to go except downhill, and downhill lay safety, so I left the rescuers to rescue themselves.

But the spearhead of this rabble was a group of real mountainmen, the famous skiing brothers Alf and Sverre Engen, former ski trooper Chuck Hibbard, and Alta Lodge manager Ted de Boer. All of them were in after-ski clothes when the alarm came. Rather than take the time to change, they elected to go as they were, lightly clad and on foot.

What that meant was plowing a furrow through the snow, with their bodies, from the parking lot to the scene of the accident.

The most difficult part of an avalanche rescue is to find a victim who lies somewhere under a mass of torn-up snow. Time is everything, for cold and suffocation are draining away the victim's life minute by minute. In the failing light, the mountain men sought for a clue. They asked one of the survivors where the party had been at the moment the avalanche struck. They were above that line of trees scattered across the avalanche path. This was a clue worth following. The boy might be lodged against one of those trees. They began to probe with wooden rods and ski poles, and they found him.

They scratched him out of the snow, using their hands like dogs digging for a bone. He was wedged in so tightly that they had to cut the wrist loops of his ski poles and the laces of his boots in order to free him. He was unconscious and badly injured but alive.

By the time we had improvised a litter out of the boy's own skis and strapped him to it, darkness had fallen. The storm had become a full blizzard. In the nightmarish glare of headlamps, six of us hoisted the litter to our shoulders and began the downhill trek. One by one, sometimes two or three at a time, the litter-bearers broke through the churned-up snow, fell, and were trampled. Other rescuers instantly took their places. Thus over a living carpet of bodies we struggled off the mountain.

The accident had a happy ending. Bruce Holm lived to ski again. But like the fires of 1910 this crisis taught a bitter lesson. I was ill-prepared to deal with this kind of enemy. I needed everything: experience, knowledge, equipment, help. Most important, it seemed to me, I had to learn how to fight avalanches. A sign reading "CLOSED AREA AVALANCHE DANGER" was not my idea of a potent weapon. In the first place, the skiers didn't like it.

Here the Forest Service had sponsored the development of a wonderful ski area and invited the public to come and enjoy it. Here I was, fencing off all the best slopes. The signs were a passive defense at best, and the skiers wouldn't obey them anyway. The only worthwhile solution would be a method of knocking the avalanches in the head so people could ski wherever they wanted to. Avalanche research in the Western Hemisphere began upon that grim night of December 27, 1945.

In the years that followed I learned how to knock avalanches in the head, and so did other Snow Rangers. The story is told in full detail in a different book.* For *The Forest Rangers,* one story will have to illustrate how the Forest Service met this new challenge. The incident is known in the avalanche profession as the Battle of Squaw Valley.

The 1960 Winter Olympic Games were awarded to Squaw Valley. The site was developed, partly on National Forest land, by two men, Alexander Cushing of New York and Wayne Poulsen of Nevada and California, who had a clearer vision than other men of what skiing was to become in the Sierra Nevadas—Spanish for "Snowy Mountains."

The California Olympic Commission faced a gigantic task, for at the time the award was made, Squaw Valley was a one-lift, one-lodge ski area at the end of a dirt road. Moreover, it had an avalanche history: main lift knocked down three times; two people killed. Two thousand was the biggest crowd Squaw Valley had known. In 1960 there would be five thousand contestants and officials alone, plus an estimated forty thousand spectators. The promoters had four years in which to get ready.

In the fall of 1956 the Games existed only on paper, but the Forest Service had pledged to the California Olympic Commission its all-out help. One of the things it did was to send a Snow Ranger. By now there were at least a

* *The Avalanche Hunters by Montgomery Atwater, Macrae Smith, 1968*

dozen men in the Forest Service capable of fighting avalanches in Squaw Valley or anywhere else. It was natural to choose out of this group the one who happened to be on a roving assignment. I was in St. Anton, Austria, when word came that I had been assigned the task of keeping an avalanche disaster off the program of the VIII Olympic Winter Games.

Hastily I canceled my plan to spend the winter of 1956–57 with the European avalanche men. I read the cablegram to the Austrian snow scientist with whom I had been spending my days.

"It is a great challenge," he said. "You are familiar with this Squaw Valley, of course?"

"Not very," I replied. "I've been there twice."

The Austrian shrugged. "Here we think it takes five years to learn a new area. Good luck, my friend."

I had plenty of time to think on the long flight home. I thought back to Alta and my first amateurish efforts to cope with its avalanches. I had found that a man on skis could cut swiftly across the slidepath at the points where it usually broke off. If the snow held together for him it would probably hold together for the other skiers. It was not an exact formula, but it worked. Avalanche men use it every day of the winter, and it even has a technical name: protective skiing. But there is a limit to how far one can trust this technique. I found that out the day I met my first hangfire avalanche. It didn't slide the moment I touched it with the cutting edges of my skis but waited a few seconds, until I was a hundred yards down the slope. I limped away from that one convinced that there must be a better method.

To an ex-infantryman, there was an obvious answer—explosives. There were some delays, due to non-skiing officials who were taken aback at the notion of turning a public recreation area into a shooting gallery. But the Forest Service was still the outfit that pioneered portable radio telephones, smoke jumpers, aerial tankers and other innovations as daring in their day as bombing avalanches. Presently Snow Rangers began stalking avalanches from the flanks and from the rear. We presented our violent calling cards and watched the cascading masses of snow waste their fury on nothing. We had made a scientific discovery, although it was some time before we got round to putting it into words. We were having too much fun blowing up avalanches and watching skiers carve their graceful patterns on slopes that had always been too dangerous before.

What we had discovered was that the avalanche is the only one of the great destructive natural forces that man can trigger at a time of his choosing. The hurricane hunter can say that the storm is moving on such and such a course at so many miles per hour. The flood forecaster can say that the water will climb to a certain level. Only an avalanche buster can make an avalanche happen at a time when it will catch no victims and before it has grown to its full power.

We were also learning about avalanches themselves. Why did snow slide on this day but not on that day, on one slope but not on another? Snow looks innocent and simple and is really complicated. It almost has a life of its own, for it is always changing, from day to day, from hour to hour, until it becomes water again flowing down to the sea. Snow Rangers spied upon it. How much snow fell, how fast? What kind of snow was it? What were the shape and size of its crystals, its weight, its temperature? How hard did the wind blow, from what direction? The instruments to measure some of the things we wanted to know didn't exist, so we invented them. There were four of us in those days: Dick Stillman in Colorado, Frank Foto in Washington, Ralph Wiese in Oregon, myself in Utah. Though we were many miles apart, it was a close-knit group. We watched, we measured, and we wrote it all down. Out of the welter of records, notes, charts and photographs, patterns of avalanche behavior emerged.

The skiers liked what we were doing. They had the whole area to play in now, not just parts of it. But more and more of them kept coming and already they had another complaint. We weren't knocking the avalanches out fast enough. Sometimes they had to wait half a day while we climbed up to the heads of the slidepaths and blew them off by hand.

There was a faster way of jamming a dose of explosives down the throat of an avalanche. All of us were ex-soldiers and had thought of it longingly: artillery. The officials

Automatic measuring devices invented by the author for avalanche research in snowfall intensity and precipitation intensity

S. Forest Service

crossed their fingers and rapped on wood again and told told us to go ahead. In the early 1950s a truck of the Utah National guard wheeled up to Alta trailing behind it a long-barreled cannon. It was an ancient weapon, of World War I vintage, but the French 75 was the finest field artillery piece of its time. The gunners trained its sinister muzzle on the mountain. Then they blew their 75mm trumpet, and the avalanches came tumbling down.

The homeward flight over the Atlantic from Austria was above clouds. With very little effort of the imagination I could think of them as snowfields. I remembered the Hard Winter of 1948–49 when all nine miles of Little Cottonwood Canyon were choked with avalanches. I remembered the Winter of the Big Snow, 1951–52, when there were over eight hundred inches—almost seventy feet—of snowfall. And I remembered the Winter of the Bad Snow, 1950–51. That year gun crews and Snow Rangers fought a pitched battle of artillery, bombs and skis against avalanches; and they won it, too. I wondered what Squaw Valley and 1960 would be like. For the next three years it would be my task to survey the battlefield, make the plans, and gather the equipment. But when the day of the Olympics came I would not be alone.

I reached Squaw Valley in January, 1957, and spent the rest of that winter listening and looking. One of the first things an avalanche buster learns is not to jump to any hasty conclusions. I rode the main lift and looked at the avalanche barrier that guarded the top terminal. The barrier was a wall of earth and rock ten feet high, set at an angle so as to divert the avalanche away from the lift. I peered up at the Headwall looming over lift and barrier, a short slope as avalanche paths are reckoned but bulging with snow. I thought that conquering this one avalanche might be my most important job. Twice it had overwhelmed the lift, and it had taken one life. It was a freak avalanche that obeyed none of the rules. I thought the barrier was too low. It would turn one slide, maybe two. After that the barrier would be filled to the brim. A third big avalanche would ride right over the lake of snow and into the terminal.

I climbed Squaw Peak to eye the awesome battlement of cornices around its summit. A cornice is a snowdrift. Wind-formed, it takes the shape of a breaking wave. As the winds blow and the snows fall, it leans ever farther out over the slope. Eventually the wave breaks—no man can predict exactly when—and bombards the mountainside with blocks of hardened snow weighing tons apiece. On Squaw Peak the cornices were forty feet high.

No one but a few cross-country skiers had carved turns on the forty-degree slopes under the Peak. By 1960 a lift would climb it. The bowl beneath those cornices would swarm with course preparation crews, officials, contestants, and spectators. The Men's Downhill, premier alpine race of the Winter Olympics, would start on the summit, curve under the cornices and plunge two miles to the valley floor. In my notebook I wrote that no less than seven major slide-paths intersected the racecourse.

On March 5, the Headwall stirred like a half-awakened giant and brushed six skiers out of its way as if they had been so many flies. Some were bruised; some lost their equipment. None, luckily, were killed. It had the earmarks of a personal challenge.

My freshman year at Squaw Valley was rather mild. I could have used another like it while I finished learning my way around, but this was not to be. 1957–58, the winter of the April Fool Blizzard, began early and lasted late. Ready or not, there was going to be a duel with the Headwall. This monstrous avalanche runs big only once a winter, so Squaw Valley people told me. I had strong reservations about that piece of advice. There was no physical reason why the Headwall should not slide half a dozen times, as often as it could collect enough snow. Because of its shape

Author aims French 75 at Mt. Superior in the first such use of artillery in the Western Hemisphere

and orientation, collecting snow was something the Headwall did better than any slope I had seen.

On my part, I had a new weapon—one coveted for years: a 75mm recoilless rifle. I had seen one for the first time just before leaving Alta. It was a far cry from the French 75. That original weapon of the Snow Rangers weighed over half a ton. It took four men to wrestle it into position on a hard surface, and it took a truck to move it from one place to another, on a good road. The recoilless rifle, that jewel of the weaponmaker's art, earned its reputation in the Korean War as the infantryman's personal artillery. It weighs just a hundred and sixty-seven pounds. Two men can take it anywhere, on a lift, on a toboggan, or on skis.

Yet it hurls the same size shell as a conventional 75, and with great accuracy.

The stunning crash of the recoilless quickly became a familiar sound in Squaw Valley as Ski Patrol Chief John Mortizia and I fought it out with the Headwall. Storm after storm, and sometimes in between, we plastered that avalanche with shellfire, and sullenly it yielded its snow, a little at a time. This was exactly what we wanted—many harmless small avalanches instead of the one blockbuster that the Headwall was in the habit of delivering.

The Headwall gave notice that it wasn't going to roll over and play dead for us. Twice it fired back at us with avalanches big enough to test the barrier. The diversion wall turned both of them, but John and I had cause to worry. The barrier was almost full. If a third big slide came it would be the fatal pattern that wasn't supposed to happen.

March is traditionally a month for heavy snowfall in the Sierras, and 1958 was no exception. Over a hundred inches of snow fell, most of them in one great storm that rolled in from the Pacific toward the end of the month. There was only one thing to do: keep on shooting; keep on bringing the Headwall down a little at a time. We fired by instruments at an enemy hidden in the storm. As fast as we shot it down the wind piled more snow on that bulging slope. As the storm was ending, the last avalanche came down. It was too big. It went over the barrier. But when we dug the lift out we found it undamaged, barring a few wrecked chairs.

We were jubilant. It was April now. No more storms; no more avalanches. We'd whipped the Headwall. I left Squaw Valley to take care of other business. Then the April Fool Blizzard struck. It took everyone by surprise—a light spring storm that wasn't expected to amount to anything but which unloaded nearly a hundred inches of snow without stopping. Highways closed and railroad trains stopped running. Aircraft could move, flying above the storm, but they could not land at any airfield within reach of Squaw Valley. I was stranded two hundred miles away, and the rifle sat voiceless on its tower.

I kept in touch with John Mortizia by telephone. Squaw Valley was snowbound. There were no skiers, but there was a crew at the top terminal trying to keep the main lift shoveled out. I urged John to bring his men in. They came, and half an hour later the place where they had been working was under thirty feet of avalanche snow. A telephone call had saved three lives, but nothing could save the lift. The terminal was demolished.

By the winter of 1958–59 the VIII Olympic Winter Games were no longer a paper project. Buildings were sprouting all over the valley floor—the Ice Arena, dormitories for the athletes, shelters for the spectators, offices, workshops and storehouses. New lifts climbed the mountains, to KT-22, Papoose and Squaw Peak. The avalanche barrier for the main lift had been doubled in size. Dick Stillman locked the doors of his avalanche research station at Berthoud Pass, Colorado, and came to join me. We had our own little army, six of the best mountainmen in Squaw Valley, whom we proceeded to train in Forest Service avalanche control technique. It was our last chance for practice, and not all practice either. The Games officials had scheduled a dress rehearsal of the Olympics in March.

Dick and I made our plan. The situation was unlike anything we had faced before. In an ordinary ski area, under ordinary circumstances, we could close all or part of the slopes if we weren't quite sure of the snow. There was another side to it. We were both research men, and no one could be more conscious than we were of the fact that the mysteries of snow were far from solved. At Squaw Valley there could be no allowance for lack of knowledge, and we couldn't close any slope, except in the most dire emergency. For the dress rehearsal, as for the Games themselves, there were to be six races on four different mountains. Day after day the course preparation crews had to be out there tramping the snow into a smooth, hard avenue of white that would stand up under the bite of racing skis. Communications crews had to string wire; course setters had to lay out their flag patterns; and in addition there were course police, TV cameramen, ski patrolmen, officials, coaches training their teams, spectators watching them, and skiers (Squaw Valley was still a public recreation area). Every one of these people had his valid reason to be on the mountain, and it was the mission of the Snow Rangers to see that not one of them got ambushed by an avalanche.

As Dick and I saw it, our problem was simple: win a slugging match with winter, no holds barred, no quarter asked or given. And so we planned it. In round numbers, there were thirty avalanche paths in Squaw Valley that could make trouble, distributed over three square miles of mountains. We would shoot every one of them to pieces, every day. Somewhat startled no doubt, but ever the friend and ally of avalanche fighters, the U.S. Army opened the doors of its arsenals and said, "Help yourselves." We took four 75mm and two 105mm recoilless rifles and vanloads of ammunition, bangalore torpedos and demolitions powder.

A week before the dress rehearsal, we got our comeuppance. The mild winter turned vicious. It was almost as if the weather had been toying with us, letting us do our puny best. Now it was going to show us who was boss of the mountain. Sixty inches of snow fell, followed by six inches of rain. Squaw Valley turned into a snow quagmire. Then it snowed another forty inches, a new load of twenty pounds pressing on every square foot of the waterlogged jellylike mess clinging to the slopes.

In the minds of the Snow Rangers, there was no doubt as to who was boss of the mountain. Nature was, with more latent power locked up in a single cloud than in man's most ferocious nuclear bomb. We called in everyone—course preparation men, communications people, officials,

Forest Service Trainees watch dust cloud at Alta, Utah

contestants, spectators, skiers—everyone but the avalanche fighters. We couldn't defeat Nature if she chose to have such a paroxysm, but maybe we could bend her a little, enough to save the lifts. We fired our ice-encrusted rifles. Shells plunged through the storm to commit noisy suicide in the hearts of the avalanches. From a safe perch on mov-

U. S. Forest Service

Avalanche Forecaster using penetrometer to determine snow density in Berthoud Pass Ski Area

ing lifts, we dropped bombs. Blasting teams inched delicately down ridges, blowing avalanches ahead of them. Bangalore torpedos sliced off cornices by the linear mile. In places the slides were coming off fourteen feet deep, right down to bare earth.

The battle had its lighter moments. A news photographer asked to go on an avalanche hunt. Nothing else was happening in soggy, slushy, stormwracked Squaw Valley and news photographers are a hardy tribe. We took him along. When the expedition was over he spoke a sentence that avalanche busters treasure: "Whatever they pay you idiots, it isn't enough." Out of his camera came a classic photograph of embattled Snow Rangers which has gone round the world.

The storm ended and the sun shone again on Squaw Valley. In a few days' time we had fired over a hundred rounds from our rifles; we had touched off over a ton of explosives. With the rubble of avalanches strewn everywhere, the place looked like what it was, a battlefield. But no building was destroyed, no lift went down, no human being lost his life. The dress rehearsal of the Games went onstage on schedule.

To the Snow Rangers, the VIII Olympic Winter Games were an anticlimax, with mild weather. Casually we skied, shot and blasted down a mere hundred and sixty avalanches. A newsman wrote: "It is the Snow Rangers' intention that not a pound of unstabilized snow will be left in Squaw Valley the day the Olympics open."

Forest officers studying terrain in Mill Creek for its suitability for skiing

U. S. Forest Service

His words were prophetic. On that day in February 1960, thousands of spectators massed in the valley and millions more watching television screens witnessed the dramatic moment during the Opening Ceremonies when storm clouds broke and the sun shone through. Up in those clouds the Snow Rangers were killing the last avalanche in Squaw Valley.

The VIII Olympic Winter Games avalanche control program was an extraordinary effort unlikely to be duplicated. But it was duplicated, six winters later at the World Ski Championships in Portillo, Chile. That story is told in *The Avalanche Hunters.* What 1960 proved was that avalanche control had come a long way in the fifteen years since an ex-ski trooper reported for duty at Alta. Today many people besides skiers benefit from the work that began there in 1945. Practically every form of human activity is pushing up into the highlands—transportation, communications, recreation, industry. All-year highways funnel a river of trucks, buses and private cars through passes that formerly were closed six months out of the year; microwave stations flash their electronic beams from mountaintop to mountaintop; at opposite ends of the hemisphere, a mine in Chile and a mine in British Columbia will soon enrich the world with their treasure—all because Snow Rangers of the Forest Service learned to tame avalanches.

10
Wilderness

Practically everyone has a yearning for the wilderness, even those of us who have never seen a wilderness and possibly never will, except in pictures or the pages of a book. This longing is a difficult thing to explain. Perhaps the instinct is inherited. A dog turns round and round, instinctively going through the motions of making a nest in the grass as his forbears did ages ago, before he lies down on the rug in front of the fireplace. In the same way a man's nerves hear the echoes of a past when all the world was a wilderness and he was free to go or stay wherever he chose.

To some men the call of the wilderness is irresistible. They become the explorers and devote their lives to finding out what's on the other side of the mountain. Such men, rocket-borne, leap into the greatest wilderness of all, outer space. In our earthbound wildernesses, such men become the game wardens, guides and Forest Rangers. Most of us must be content to taste the peace and quiet of the wilderness on a vacation.

The wilderness is not either peaceful or quiet entirely; it is full of sound and movement if one keeps quiet enough to observe them. The battle for survival goes on. Water gurgles over rocks, the wind whispers through aspen leaves;

. S. Forest Service

Beyond the farthest campfire

Wilderness camp

. S. Forest Service

a mother grouse clucks to her chicks; an elk bugles his challenge; a coyote cries his lonesome lament. In the wilds, all sounds are lower and softer. The thump of a pine cone falling to earth seems louder than a pistol shot.

Everyone who has spent time in the wilderness has his own special memory of it. This one is my own. It was a July afternoon, cloudless and hot. No breeze stirred. Alone and on foot I discovered a lake. At least it was far off the beaten path, and it didn't show on my map. If any man had been here before, he hadn't left his rusting tin cans, his broken glass, or the ashes of his campfire.

On the far side, the forest marched down a steep slope to the water's edge. Behind the forest, a semicircle of cliffs completed the illusion of a stage setting. I approached the lake across a meadow bright with Indian paintbrush, bluebells and shooting stars. As I crashed along in clumsy man-fashion, no insect chirped, no bird called, no fish jumped. Except for me the place seemed to be deserted.

I knew better. Like a crossroads store, a lake is a meeting place of the wilderness. I immediately named it Beaver Lodge Lake. Near the outlet was one of the largest beaver houses I had ever seen, a conical mound of mud and sticks well over six feet high. A beaver may not be the neatest carpenter in the world, but he builds solidly. Within his fort he sleeps immune to heat or cold or enemies.

I climbed the sloping wall and made a comfortable seat for myself on top. There I was, in plain view of hundreds of pairs of wilderness eyes. But it is not necessary to hide in order to observe the wild folk. I sat still. Within ten minutes there were rustlings and chirpings in the grass as the smaller creatures decided it was safe to come out. In half an hour the community had either forgotten about me or decided that my intentions were peaceful. Marmots whistled from the rockslides. Squirrels performed acrobatics in the trees. Trout jumped exuberantly all the way out of

U. S. Forest Service

Canada geese flying over a Northern Eastern Region Forest, Wisconsin

water as they struck at the insects hovering just above the surface. A doe tiptoed down to the lake, drank, caught a startled glimpse of me over one shoulder and bounced away, but at no more than half speed. From the shelter of a rocky point a flotilla of ducks sailed out into open water.

There was only one flaw in this picture of wilderness peace and goodwill. The owner of the house which I had chosen as a vantage point was having none of it. He had been among the first to show himself, a giant of a beaver. No doubt my scramble up to the top of his citadel had wakened the old warrior from his afternoon siesta. I was aware of him first as a pair of shoe-button eyes and a larger button nose which simply materialized offshore, without a ripple. With a safe ten feet of water between us he glared and blew threateningly through his nose. That blowing noise and the little chirp which is a love cry are the only sounds a beaver can make, except with his tail.

His meaning was plain. I was a trespasser on his house, at his lake. Would I kindly get lost?

Using only fingers and wrist, I flipped a bit of sun-dried mud at him. The response was impressive. Slaptail Bill hoisted his stern out of water, flourished his paddle-shaped tail in the air, and brought it down with a thwack that resounded all over the lake.

I wished I hadn't tormented the old fellow. Surely the racket he was making would send every creature within a mile back into hiding. But apparently the citizens of this wilderness community were used to the terrible-tempered beaver. The trout went on jumping. Mice scurried in the grass. The fleet of ducks executed column right and headed in toward Bill's cove.

Bill continued to glare at me. "One more false move," he seemed to say, "And I'll *really* let you have it."

He wasn't aware of the ducks until they sailed right past him. It was then that I witnessed the act that keeps this day bright in my memory. It is not uncommon for wild creatures to warn each other of danger. Squirrels swear at a coyote ghosting through the forest. Gophers whistle at the shadow of a cruising hawk. A band of elk always has a sentinel. But I had never seen or heard of such a performance as the old beaver put on.

He deliberately tried to drive those ducks away. He swam up and down in front of them. He blew through his nose. He slapped ferociously with his tail. Always he kept himself between man and ducks. It was an act of heroism, for the beaver tribe has every reason to fear the man tribe. On their part, the ducks either had not noticed me or had decided that I was harmless. They quacked amiably to the frantic old beaver. Apparently they were used to his tantrums. When he made a dash directly at them, they opened ranks and let him through. But they had business in his cove, and determinedly they steered for the shore.

It was too much for me. I laughed. This raucous man-

U. S. Forest Service

Devastation by a colony of beaver in New Mexico. All the felled timber on the hillside and in the background is their handiwork.

Beaver dam on Mission Creek demonstration watershed (Washington)

J. S. Forest Service

U. S. Forest Service
Robert Marshall, Chief, Division of Recreation and Lands, National Forest Administration, U. S. Forest Service (May 1937–November 1939)

sound ended the game. With startled cries the ducks took off and flew to the far side of the lake. I departed, and Slaptail sped my retreat with whole salvos of tail-thwacks. Let's hope he's still there, building his house ever higher and thicker, guarding his ducks and all his other subjects in that unspoiled wilderness.

If he is, and if there are many other unspoiled wildernesses in our country, it is due mostly to one man. Robert Marshall was an unusual person, a top official of the Forest Service who could seldom be found in his office in Washington. He spent most of his life roaming the back country. Usually he traveled alone, for the simple reason that no

'. S. Forest Service

The Chinese Wall, Bob Marshall Wilderness

one could keep up with him. Among foresters he was labeled a trailburner. He could outwalk any man and most horses.

Perhaps the voice of the wilderness called to him more loudly and more clearly than to other men. He was not alone in loving it and wanting to see it preserved; many men had that idea. But the type of person who prefers the solitude of the back country is generally shy and reserved, not the kind to go into the marketplace and sell an idea. Bob Marshall was equally at home in the wilderness or in a crowd. He had the imagination to see how a wilderness can exist in the age of superhighways, and the eloquence to make his dream come true. The Wilderness areas of the

U. S. Forest Service

Wilderness Patrolman William James ready to start out on a patrol trip in the Sawtooth Primitive Area

Snow Ranger Ed Rockwell pauses to admire the wild, alpine beauty along the trail to the world-famous Minarets of the Sierras

U. S. Forest Service

Forest Service are the product of many hands and many minds, but more than to any other individual we owe them to Bob Marshall.

Nowhere else in the world is there anything exactly like a Forest Service Wilderness. It is not, for instance, a National Park, which is a sort of outdoor museum. There are close restrictions on where one can go and what one can do in a Park. A Wilderness is a carefully chosen part of a National Forest, usually a remote part. Thus the first line of defense of a Wilderness against encroaching civilization is distance. In a Wilderness there are many trails but no roads, hotels, motels, gas stations or stores; no habitations either, except for the lookout towers and ranger stations.

Cooking and stove set-up in Spanish Peaks Wild Area of Montana

S. Forest Service

There may be airfields, but they are for the use of firefighters. The only price of admission to a Wilderness is one's personal desire to get into the back country and the effort it takes to get there, on foot, on horseback or by boat. A person may camp where he pleases, hike over any trail, fish and hunt in the proper seasons—do almost anything except set the place on fire.

The boundaries and the number change from time to time, but there are today about seventy-five Wildernesses containing about fourteen million acres. More and more people are discovering them, but they are so big and so remote that meetings are few. I remember a trip into the Jim Bridger Wilderness of Wyoming, named for the famous frontier mountainman. My companion and I saw no one else from the moment we stepped off the mainline trail to do a little bushwacking until we came out ten days later.

This feeling of separation from the rest of mankind, of freedom and independence, is part of the spell of a Wilderness. The other part, of course, is magnificent scenery, streams where the trout practically come out on the bank to greet an angler, where that master ventriloquist, the coyote, bounces his voice off the mountainsides until he sounds like a dozen. A hawk rides the aircurrents on silent wings; the smell of pine and curing grass comes with the up-canyon breeze of morning. These are the sights, sounds and smells of the Wilderness, yours for the taking. But the visitor is not quite alone. The setting sun flashes on the glass walls of a distant lookout tower; overhead sounds the faint buzz of a patrol plane. The Forest Rangers are with you.

U. S. Forest Service

The ranger must travel on skis in winter to reach back country stations. This ranger is hauling several boxes of powder to blow out an ice jam on Big Smokey River (Sawtooth National Forest, Idaho).

U. S. Forest Service

Penny, young daughter of Powell Ranger District Assistant John Watson, with a white-tailed deer fawn

I have mentioned one of the greatest of the Wilderness Areas several times in this book, that one which lies along the South Fork of the Flathead River in northern Montana. Very fittingly it is named for Bob Marshall, more than any other man the father of the Wilderness idea.

11

Do You Want to Be a Forest Ranger?

The easy way to help a person answer that question would be simply to set forth the pay scale and fringe benefits, describe the qualifications, and list the addresses to which one can write for information.

It seems to me, however, that there is a good deal more to be said. In the first place, why should one want to be a Forest Ranger? The education and training are very stiff, there is a great deal of responsibility, and a ranger works very hard. Anyone with the same qualifications can probably get higher pay in private industry, for less effort. There is no one-sentence answer to the dilemma. The closest I can come is to say that it is a good way to live. As I stated in the foreword to this book, no one is likely to become wealthy working for the Forest Service, but he will never be bored. That last I can guarantee, from personal experience.

A reader of *The Forest Rangers* can scarcely have missed the fact that the author of it is a great admirer of the Forest Service, its principles and its works. I saw it from practically every viewpoint: as an outsider, as a friend and neighbor, as a part-time employee, as a career forester and as a self-appointed historian in both fact and fiction. Perhaps it will do something to clear up this business about wanting to be a forester if I tell very briefly how I became one.

I had the good fortune of growing up in a National Forest, and by "growing up" I mean that period when one changes from a youth to an adult and gives serious thought to one's self and future. Some of the Forest Service's greatest achievements took place during this period and I saw them. I learned to hunt and fish in the ghost forests of 1910, when the most a ranger could do was bring his men out alive. And I was at the Battle of Camptonville when rangers challenged a fire storm head-on and knocked it back on its heels. Foresters taught me to read map and compass, to pack Decker fashion. I learned a lot about conservation—and about skiing—on a winter game patrol.

Probably the best preparation for a career in the Forest Service is growing up with it. I had often thought of joining the Forest Service. I liked the people and the way they lived. I admired what they were doing. But there was an obstacle—no college degree in forestry. This meant that I could never climb higher on the career ladder than the rung below Forest Ranger. My conclusion was that I'd better stay with my original plan of becoming an author. The Forest Service should make a good subject. And it did.

A great many of my personal plans were drastically remodeled by World War II. Like many others, I came out of the Armed Forces needing, first of all, a job. On its part, the Forest Service had a need for a new kind of forester, the snow ranger, as result of the rapid development of skiing. A soldier's needing a job coincided with a ski area's needing an avalanche man. Thus I became a career "forester," as a specialist.

It was a good career; it all comes back to that. Being a

forester is a good way to live if one isn't plagued by a yearning for wealth. There are the surroundings, which are unfailingly beautiful. There is adventure—sometimes a little too much of it. And there is the feeling of doing something worth while, for others as well as for oneself. I don't intend, by the way, to give the impression that a forester should be without ambition and drive. The career ladder whose foot rests on the doorstep of a raw recruit has its top rung in the office of the Chief of the Forest Service. Every Chief in the Service's history has climbed it.

In Ouschita National Forest, Arkansas, Jim Burner, forestry technician, instructs Job Corps enrollee Floyd Castle of Coeburn, Virginia, about the operation of the D7 Dozer

U. S. Forest Service

Few people considering a forestry career can grow up with it as I did. Neither is it necessary. What is necessary is to plan for it. A modern Forest Ranger must have that college degree in forestry, a program that begins in grammar school. He must also have experience and training. Joining the Boy Scouts is a good way to start learning mountain and woodscraft.

Every summer the Forest Service hires a large number of seasonal workers for a great variety of jobs. These summer foresters repair roads, trails and bridges and build new ones. They work on bug and erosion control, reforestation and range improvement. They are lookouts, smoke jumpers, equipment operators—the list could go on and on. Forestry students have a prior claim to these jobs, for this is the way they get the practical experience to go with their formal education. The counterpart of a college professor in lecture hall or laboratory is a Forest Service project leader in the field. Each has an essential part in the making of a forester. One other thing happens in these schools of the great outdoors. The would-be forester finds out whether or not he likes the forest, and whether or not the forest likes him.

The way to get one of these summer jobs is first of all to be eighteen years old. The Forest Service will hire no one younger. It is not necessary to apply in person, but by all means one should do so if there is a Forest headquarters near enough. And don't wait until school lets out, either. While there are many jobs, the competition is keen, and the very worst time to approach a ranger is in the middle of the spring rush.

Schools of forestry operate placement bureaus for their students. The Forest Service itself puts out pamphlets and instruction sheets to help anyone interested in a forestry career. If there is a Forest headquarters near enough, go to it and ask for this information. If one writes for information, it is better to send the letter to a Forest Supervisor's or Regional Forester's office. Forest and Regional headquarters have personnel officers whose job includes taking care of such requests. There is a list of Forest Regions at the end of this chapter. It is also perfectly proper to write to the Chief's office: Forest Service, U.S. Department of Agriculture, Washington, D.C. 20250; Attention: Division of Personnel Management. Finally, anyone lucky enough to have a friend or relative who is or has been a forester should ask him or her to help. There is a mystique about the profession that makes every forester a recruiting agent.

Many more things could be said about preparing to become a Forest Ranger, but I think it best to keep it simple. Get the education; get the practical experience. The rest is between you and the forest.

In this chapter on a Forest Service career, the emphasis has been on the ranger, as it has been throughout the book. But as I mentioned in the second chapter there are only about eight hundred Forest Rangers in a Forest Service army of about fourteen thousand. Some of these others are "overhead"—Forest Supervisors, Regional Foresters, staff officers—but most of them are assistants, trainees and specialists. Modern forestry is a very large and complicated enterprise, and a Forest Ranger needs a lot of help.

There is scarcely a skill, craft or profession that is not represented in the Forest Service, from lawyers to aviators, from computer programers to plant biologists, from clerks to mechanics, from artists to engineers. I won't attempt to list them all but will simply end as I began by saying that the Forest Service offers lifetime careers to girls as well as to boys in a most useful and interesting profession.

Forest Regions and Regional Offices

(Note that there is no Region 7. It was recently combined with Regions 8 and 9.)

Region 1 NORTHERN
(Montana, northern Idaho, northwestern South Dakota, northeastern Washington)
Headquarters: Federal Building
Missoula, Montana 59801

Region 2 ROCKY MOUNTAIN
(Colorado, Kansas, Nebraska, South Dakota, Wyoming)
Headquarters: Federal Center
Building 85
Denver, Colorado 80225

Region 3 SOUTHWESTERN
(Arizona, New Mexico)
Headquarters: Federal Building
517 Gold Avenue, S.W.
Albuquerque, New Mexico 87101

Region 4 INTERMOUNTAIN
(Utah, southern Idaho, western Wyoming, Nevada)
Headquarters: Federal Office Building
324 Twenty-fifth Street
Ogden, Utah 84401

Region 5 CALIFORNIA
(California, Hawaii)
Headquarters: 630 Sansome Street
San Francisco, California 94111

Region 6 PACIFIC NORTHWEST
(Oregon, Washington)
Headquarters: 319 Pine Street, S.W.
P. O. Box 9623
Portland, Oregon 97208

Region 8 SOUTHERN
(Alabama, Arkansas, Florida, Georgia, Kentucky, Louisiana, Mississippi, North Carolina, South Carolina, Oklahoma, Tennessee, Texas, Virginia)
Headquarters: 50 Seventh Street, N.E.
Atlanta, Georgia 30323

Region 9 EASTERN
(Connecticut, Delaware, Illinois, Iowa, Indiana, Maine, Maryland, Massachusetts, Michigan, Minnesota, Missouri, New Hampshire, New Jersey, New York, Ohio, Pennsylvania, Rhode Island, Vermont, West Virginia, Wisconsin)
Headquarters: 633 W. Wisconsin Avenue
Milwaukee, Wisconsin 53203

Region 10 ALASKA
Headquarters: Federal Office Building
Box 1628
Juneau, Alaska 99801

Forest Careers Publications

(A list provided by the Forest Service)

Conservation Occupations—Foresters, Forestry Aids, Range Managers. U.S. Department of Labor, Bur. of Labor Statistics. 1966. 7 p. (Occupational Outlook Handbook. Bull. No. 1450–20) 260–1

Get Ready for College and Go. Office of Education. 1965. (Folder) (OE-50046) 260–2

Financing an Undergraduate Education. U.S. Department of Health, Education and Welfare, Office of Education. 1965. 21 p. (OE 55035A) 260–3

The Forest Service Engineer. Your Gateway to the Future. Describes the U.S. Forest Service—What it is, what it does, and how it operates. Includes information on the varied and interesting engineering activities offered to the graduate engineer. 1963. 32 p. (Mis. Pub. No. 841) 260–4

Opportunity for You. A Forest Service Career in Research and Administration. Forest Service. Jan. 1957. (Folder) (MP-726) 260–5

Challenge in Wood Research. 1967. 13 p. (Misc. Pub. No. 1054) 260–6

Opportunity in Forest Service Management Careers. Forest Service. 1965. 33 p. (PA No. 535) 260–7

Employment Outlook for Foresters. (Labor Dept.) 260

Forestry Schools in the United States. List of Schools Offering Accredited Courses in Forestry. 0–11 FS-9260/210

A Job with the Forest Service. 1968. Misc. Pub. No. 843. 260

So You Want to Be a Forester. 260

A Summer Job with the Forest Service. 0–29 260

The U.S. Forest Service Offers You, etc. 260

Heads of the Forest Service and Their Predecessors

Franklin B. Hough 1876–83
(*Division of Forestry established 1881*)
Nathaniel H. Egleston 1883–86
Bernard E. Fernow 1886–98
Gifford Pinchot 1898–1910 (*First Chief of the Forest Service*)
Henry S. Graves 1910–20
William B. Greeley 1920–28
Robert Y. Stuart 1928–33
Ferdinand A. Silcox 1933–39
Earle H. Clapp 1939–43
Lyle F. Watts 1943–52
Richard E. McArdle 1952–62
Edward P. Cliff 1962–

Highlights of Forest Service History

The following are taken primarily from U.S. Department of Agriculture Information Bulletin No. 83, Forest Service:

Colonial Times Forest covered the land from the Atlantic Coast to the Great Plains. As Gifford Pinchot said more than a century later, "Most people felt that the thing to do with a forest was get rid of it."
Nevertheless, in 1710, the first community forest in the United States was established at Newington, New Hampshire, and it has benefited the town for two hundred and fifty years. It is a striking example of what was to become the Forest Service principle of conservation: use, protect and maintain.

1799 The Federal Timber Purchases Act appropriated $200,000 to buy timber for naval purposes, early recognition of the need to husband timber resources.

1828 Santa Rosa, a peninsula jutting into the Bay of Pensacola in Florida, was established as our first forest experiment station. Trees were planted on this 30,000-

acre tract, fire lines were opened, brush was cleared, and selective cutting was practiced. Unfortunately the project was abandoned after two years.

1871 The Peshtigo Fire in Wisconsin burned 1,280,000 acres. Fifteen hundred persons lost their lives; many towns and settlements were destroyed.

1872 Yellowstone National Park was reserved as a "pleasuring ground," the beginning of the National Park system. Arbor Day was instituted in Nebraska to stimulate tree planting in the prairie country. Observance of Arbor Day has since spread to every state and to many foreign countries.

1873 Yale University offered lecture courses on forestry, probably the first in the United States.
Congress passed the first timber culture act, granting a homesteader a patent to 160 acres of land if he planted one-fourth of it to trees.

1876 A hundred years after the Declaration of Independence, Congress made the first move toward forest conservation on a national scale by ordering an investigation of timber resources. Dr. Franklin B. Hough became the first forestry officer in United States history. To finance his job, Congress appropriated the sum of $2,000.

1877 Carl Schurz, an immigrant from Germany, well-known scholar, statesman and forester, became Secretary of the Interior. Congress directed him to make plans for establishment of a Division of Forestry. He was an early advocate of forest management as practiced in Europe: a tree planted for every tree cut.

1881 A Division of Forestry was organized within the Department of Agriculture. It was a fact-finding agency only and had no forests under its control.

1886 Dr. Bernard E. Fernow, another German forest conservation expert, was made the first formal chief of the Division of Forestry.

1891 The birth of a real system of national forests came with an act of Congress authorizing the President to establish forest reserves on the public domain. On March 30, President Harrison created the million-and-a-quarter-acre Yellowstone Timberland Reserve in Wyoming.

By the end of his term, President Harrison had reserved thirteen million acres. There was no plan for administration or use. These lands were simply set aside.

1897 At the close of his term in office, President Cleveland added twenty million acres to the Forest Reserves. Congress passed the Act of June 4 outlining a system of organization and management for these public forests. The act authorized the hiring of administrators and opening of the reserves to use. This act, with amendments, is the basic law of the National Forest system.

1898 Gifford Pinchot became head of the Division of Forestry. At this time administration of the Forest Reserves was still in the Department of the Interior; the Division of Forestry, in the Department of Agriculture, was a fact-finding and advisory unit. Pinchot greatly expanded the Division's activities. Actual work in the field began.

The first four-year course leading to a forestry degree was offered by Cornell University. During the next five years, other universities followed suit. Forestry became a recognized profession in the United States.

1899 The value of the forest reserves for recreation was recognized by act of Congress.

1905 Sponsored by President Theodore Roosevelt and Gifford Pinchot, an act of Congress transferred the Forest Reserves from the Department of the Interior to the Department of Agriculture. The Bureau of Forestry became the Forest Service.

1906 President Roosevelt increased the Forest Reserves to a total of a hundred and seven million acres. Timber sales and grazing permits were issued for the first time.

1907 Political elements hostile to the National Forest concept succeeded in passing legislation that prohibited the President's adding to the Forest Reserves by proclamation any further lands in the States of Oregon, Washington, Idaho, Montana, Colorado and Wyoming. Before this legislation could go into effect, President Roosevelt signed proclamations that added some 15,750,000 acres to the Forest Reserves. The boundaries and area thus established, 185,000,000 acres, one-tenth the land surface of the United States, are essentially those of the National Forest system today.

The name "Forest Reserve" was changed to "National Forest." Secretary of Agriculture Wilson sent to Gifford Pinchot, first Chief of the Forest Service, his famous "greatest good of the greatest number" declaration of policy.

1908 The first forest experiment station, excepting the abortive Santa Rosa station of 1828, was established in Arizona, quickly followed by others in Colorado, Idaho, Washington, California and Utah.

1910 The Great Idaho Fire devastated two million acres of National Forest land in Idaho and Montana, plus an estimated million acres more of state and privately owned land.

The Forest Products Laboratory was established by the Forest Service in cooperation with the University of Wisconsin. Now world-famous, it was the first research organization of its kind.

1912 An act of Congress allocated ten percent of all National Forest receipts to the building of roads and trails.

1920 The Forest Service in California made a cooperative agreement with the War Department for aerial patrols during fire season.

1922 Congress made the first appropriation, ten thousand dollars, for improvement of public campgrounds on National Forests.

1924 An act of Congress authorized the Secretary of Agriculture to make cooperative agreements for fire protection of state- and privately owned forests.

1925 The aerial fire patrol was expanded to cover Oregon, Washington, Idaho and Montana, as well as California.

1928 An act of Congress directed the Forest Service to make a nationwide survey of forest resources and conditions.

1929 During the worst fire season since 1910, fire-storms again threatened the Northwest. Because of the network of roads and trails that had been built, as well as improved methods of fire-fighting, losses were held to a fraction of those suffered in 1910.

The first aerial drops of cargo were made.

1930 An act of Congress authorized a great expansion of tree planting on National Forests.

1933 The Civilian Conservation Corps was sponsored by President Franklin D. Roosevelt as a means of providing useful work for young men unable to find jobs during the business depression of that time. During the nine years of the CCC program, more than two million men enrolled. As many as half a million were employed at one time. They planted two and a quarter billion trees and did an incalculable amount of other conservation work.

1938 The New England Hurricane blew down millions of trees. The Forest Service was given the job of directing salvage operations, because of the value of the timber and in order to reduce the fire hazard. Seven hundred-million board feet of timber was harvested before rot and insects could make it useless. This was the forerunner of salvage operations in burnt or diseased forests that are standard practice today.

1939 The first smoke-jumping experiments were carried out.

1940 During the worst season on record for the number of lightning-set fires in Montana and Idaho, the smoke jumpers became operational.

1946 The first avalanche research station in the western hemisphere was set up at Alta, Utah, by the Forest Service.

1947 Congress passed a Forest Pest Control Act, recognizing that disease and insects are as destructive to forests as fire.

1949 The reforestation laws passed in 1930 were greatly improved and expanded.

1955 Fiftieth anniversary of the establishment of the National Forests and the Forest Service.

1956 Annual National Forest income from the sale of timber, grazing rights and other uses passed the hundred-million-dollar mark.

The first practical use of aerial tankers for fire fighting was demonstrated in California.

1957 The Forest Service inaugurated Operation Outdoors, a five-year program to improve and expand recreation facilities in National Forests.

1959 During a climactic fire season in California, firefighters won the Battle of Camptonville.

1960 During the term of President Eisenhower, Congress passed an act confirming the five-fold nature of the forest resources: wood, water, forage, wildlife and recreation. The act stated the principle of multiple use.

Forest Service Snow Rangers provided avalanche protection for the VIII Olympic Winter Games at Squaw Valley, California.

National Grasslands, nearly four million acres of dry grazing lands in the western Plains, were added to the National Forest System.

1964 A nationwide review of standing timber conducted by Forest Service showed that for the first time annual growth exceeded annual cut, but average quality continued to decline.

Congress passed the Wilderness Act, classifying over nine million acres of National Forest land as Wilderness and five and a half million acres as Primitive Areas, endorsing long-standing Forest Service policy to keep such areas in their natural state.

1966 The first full year of operation for the Job Corps program on National Forests saw 6,500 young men from deprived backgrounds successfully complete a program of self-improvement and conservation work.

On March 1, the U.S. Treasury received the two-billionth dollar of National Forest receipts.

1967 The northern Rockies and the Pacific Northwest had one of the worst fire seasons on record. For the first time, all National Forests in these regions were closed to users and visitors for more than a month. Despite two months of very high to extreme fire danger, from mid-July to mid-September, very few fires reached disaster size, thanks to long advance planning and close cooperation and coordination of many Federal, state, and private agencies with trained men and the latest equipment. The largest fire-fighting air armada to date—over two hundred airplanes and helicopters—was used to ferry thousands of firefighters, hundreds of tons of cargo, and to drop millions of gallons of chemical fire retardant.

Index

Montgomery M. Atwater

Born in Baker City, Oregon, in 1904, and graduated from Harvard College in 1926, Monty Atwater enjoyed a varied career as football coach, cattle rancher, guide, trapper, and free-lance author. He served as winter warfare instructor and advisor during World War II, participated in the Kiska Campaign and the Battle of the Bulge, then retired from the army because of physical disability, with the rank of Captain.

Joining the U.S. Forest Service in 1946, he distinguished himself as a specialist in avalanche research and control and winter recreation, working until 1956 at Alta, Utah, where he founded the first avalanche research center in the Western Hemisphere. Snow and avalanche research at Alta eventually produced all the modern techniques of avalanche control now in use throughout the Americas. He planned and supervised avalanche control for the Olympic Winter Games of 1960 and developed the "avalauncher," the first avalanche control weapon designed specifically for the purpose. He retired from the U.S. Forest Service in 1964 and since that time has been a consultant on snow and avalanche problems for ski areas and industrial concerns from South America to Alaska.

At present, Mr. Atwater lives in Squaw Valley, California, where he has a small research laboratory of his own. He has two sons, Robert and Montgomery the second.

He is the author of technical and popular articles in nationally circulated publications, as well as *The Avalanche Handbook,* the first comprehensive field manual of avalanche control, and has also written numerous adventure novels for young people.